42 Plus 1
(The Secret That Science Dares Not Whisper)

by

MOL SMITH

ONVIEW BOOKS

42 Plus 1

(The Secret That Science Dares Not Whisper)

Published by Onview.net Ltd
In association with www.createspace.com
2015

Onview.net Ltd. Registered Office:
Frilford Mead, Kingston Road, Frilford. Abingdon.
Oxfordshire. OX13 5NX England

www.onview.net
www.42plus1.net

Many thanks to Lesley Evans for proof reading my book.

First Published 2015 by (Onview Books) Onview.net Ltd.
In association with www.createspace.com

A CIP catalogue record for this book is available.
ISBN-13: 978-1506022895
ISBN-10: 1506022898

42 Plus 1
(The Secret That Science Dares Not Whisper)

Mol Smith

Dedicated to my brother, Dennis, for all our talks about such things as little boys; and to my Lesley for tolerating all my lectures about the same stuff; and to all who question life from birth to death, and all who refuse to become conditioned by faith or science; for all who forever journey through life with eyes wide open and with ever-curious minds seeking truth and wisdom.

CONTENTS

Please note: This book also has internet references. Some are quite long to type in. A complete set of simplified references can be found on the associated and supporting web site for this work. Please visit:

www.42plus1.net

Scientific research in the past few decades has led to a profound change regarding the nature of reality.

Things are not what they appear to be.

There are powerful implications which science dares not comment on publicly.

New discoveries point to a startling set of conclusions.

Their consequence indicate something extraordinary which affects all humanity.

It is something almost impossible to believe, but the odds are wildly in favour of it being astonishingly true!

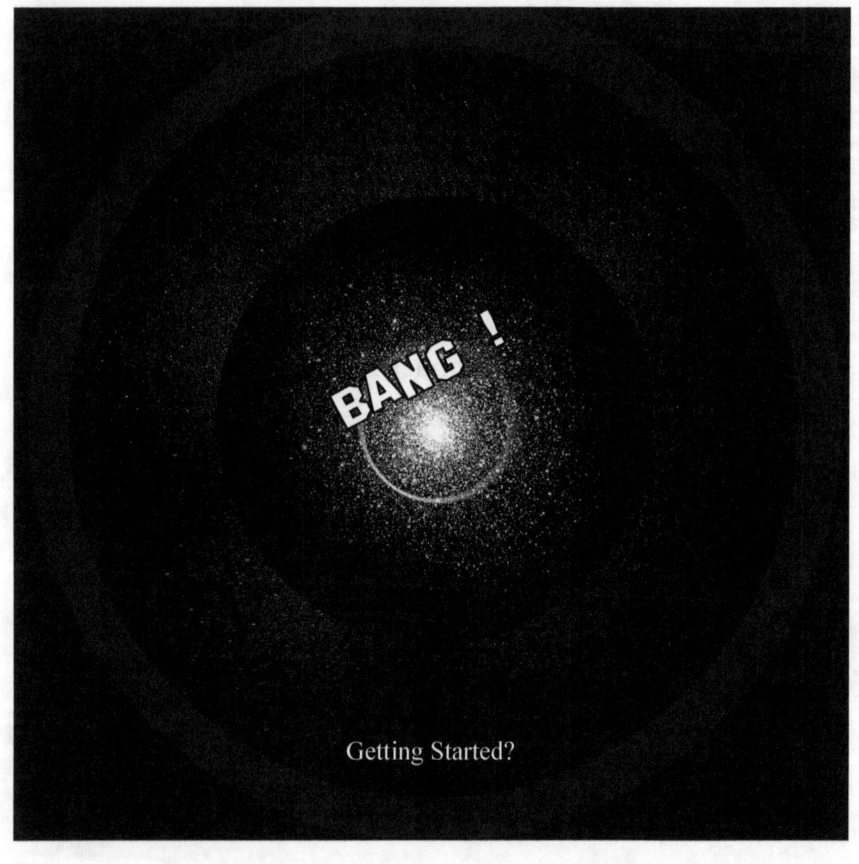

Getting Started?

Introduction

This is not the first book I've written exploring the question of life. My first book, 42—The Answer To Life, The Universe, And Everything, was published in 2007, and still sells copies every month. That work was in many ways a prelude to this. It was a collection of my 'first written thoughts'. A few years on and here I am again attempting to put something nearly impossible to perceive into a comprehensive understanding. Scientific research has moved on too, and I've stayed with it. I am now ready to put forward arguments, insights, and discoveries to support ideas which challenge much of conventional thinking in our world. It is never a good thing to challenge conventional wisdom without offering evidential or reasonable information to support the challenge. Much of what I need to put forward isn't necessarily saying this or that is wrong. It's more about looking at the implications of knowledge and theories which science dictates are probably true. It is also about redefining ideas of faith in a new light—ideas which suggest religion at its core, is more a branch of philosophy, and thus it is another tool to help make sense of human existence.

When I searched the various forums on the web to consider feedback to my first book, I was quite shocked to discover atheist site forums were often filled with nasty, hateful, and mindless cynicism and comment. Elsewhere, comments were mixed and balanced. Although there was one exception which appeared as a reader review of my book on Amazon. The comment was, "As a Christian, I found this book a complete waste of time and a total misunderstanding of God's purpose..." or something like that.

This book is not here to replace any faith with a new one. I am not a religious man. Why should I write a religious book? I am a man of science, but not the dogmatic science which currently seems to have evolved over the last few decades. It is a sad fact that what some of the finest minds in history have probed and discovered, should all be lumped together into a 'safe' kind of politic to distance itself from spiritual thinking. Most educated people already realise all faiths are each attached to an historical legacy of culture, and archaic stories, and yet I believe core values can be distilled from each to form a sense of spiritual intent to our lives. Before science, there was only imagination. Stories of sufferance along with racial history, the handing down of cultural record, and the re-shaping of moral guides into laws to control the masses by the few, all contaminated what at heart was a basic philosophy. People realised they had never made themselves. Concepts grew up to try and answer the questions of who we are, why we are, do we have purpose—and if we do, what is it?

I live in the 21st century. I can see clearly what we have learnt about where we live, and what we have discovered about our position within the universe. I appreciate science is the tool which provides this knowledge. So, I wish to say straight away, I'm not here to preach any idealism or convert anyone who has already answered their own questions of life to any new idea. How can I? Their minds are already shut tightly and they shouldn't really be buying this book. My work is for people who have not completely closed their minds. They may have a faith. They might believe in science, be agnostics, atheists, religious, spiritual, etc, or simply be people who don't know. Most 'normal' people live their lives with a mixed set of notions and they seem quite happy doing that.

I suspect those readers with closed minds who buy this book do so hoping what I say supports their specific view on the meaning of life. Alas, if you are one of these, you will have probably wasted your time. My views will not strengthen any existing faith or idealism over another, if indeed it has any influence on beliefs per se. They might, but I do not write to try and achieve that. I'm not spending months and years of personal research to bang away at a keyboard merely to repeat what other works all seem to say about life, death, the universe and everything in their own diverse and often opposing disciplines. My goal is to illuminate how we should reconsider the reality we find ourselves part of in greater depth. In doing so, I strive to offer a fresh new look at what we all ultimately consider to be the three poorly answered life questions: who am I? Why am I here? What happens when I die?

I believe science has discovered greater insights to suggest probable answers to these questions, yet fails to find a considerate and philosophical voice to tell everyone what it might mean for them. It would not be much of an exploration to find answers to these things if I simply wrote a two page book with one page answering each question. And who would consider the answers convincing unless they have had the opportunity to absorb and ponder the knowledge and perceived wisdom garnered so far? Some questions do not have a simple 'it is this' or 'that' answer. An understanding of why the answer is what it is, needs to be shared too.

It seems we live in an ever faster changing time. It brings with it a new problem. Who gets time to think? Before the internet, before mobile phones, before the current generation, before the increased and ever over-simplification of information to fit all-comers as part of a new global push to reduce life into a profit-and-loss financial endeavour, there was more time to ponder; time to think; time to consider; time to reflect on the wonder of being alive. It's gone. People today are bombarded with information bites. They are channelled into pre-described categories of simple thought, quickly satisfied desires, and unimaginative education. Want a job? Go this way. Want to know something? This is the quick answer. Google it.

I created one of the most successful science-based sites on the internet. I didn't do it alone. It took hundreds of people, and a handful of dedicated like-minded helpers to achieve it, plus twenty-odd years of unpaid work. Everyone involved had different views on life. Some of us had/have very polarised views, but avoid argument. We respect each other. We share something in common. We wish to share information, good information. We do it knowing and watching the birth and development of replicated wrong information. We realise our site is visited by millions of people around the world who come looking either for a quick answer for their homework, or to satisfy a sudden and novel wonder which triggered them into action. But we also realise the site is visited by people who know as much as us, if not more—people who come looking not just to flit... but to 'consider'.

This book is for similar people—those wishing to think about things properly; people from all cultures who wonder if there is a meaning to life, and if so, what it might be. I think I should try and draw you into my world right away, perhaps by starting with a simple truth as taught through science education in every school in the world: our reality, and everything in it, is made up of tiny parts called atoms. These are structures so small, only expensive and highly developed scientific tools can perceive them. Each atom has a nucleus and an outer set of orbiting electrons, a bit like a solar system.

Would you agree this is a nice simplified piece of science taught to us in our school days—a concept most people seem to accept as a truth, now, here in the early 21st century? Yes? Good. Well we agree on that one then, except... well me, actually. I don't think this is how it is at all. What this model sets out to do is to put something very sophisticated, elegant, and widely exotic into a simplified working notion and concept which fits our human perception of a macro-sized world. And although it succeeds in being a good model, helping us make sense of many actual working processes in many disciplines of science, the model itself is flawed and fuzzy. I don't expect you to take my word for it and later on I will put forward evidence to prove the model of atoms, despite being successful in our ability to manipulate our physical world in clever ways, is over simplistic. Thinking of matter only as tiny lumps of something solid, is one of the reasons why our minds cannot grasp more exotic perspectives on a universe which is far from being a simple mechanistic machine composed of atoms and energy packets. But, as I said, more about this later.

I expect a few readers will close the book now and chuck it to one side. They are the ones who thought they had open minds, but in fact find it so difficult to take seriously an author who apparently refutes something they believe, and have believed in, since childhood—something that everyday proves to be a good working model for every aspect of biology, science, and technology. The thing most of us are taught in school is this:

Atoms (according to the model) are composed mostly of empty space. In fact, the magnitude of space compared to so-called solid material (protons, neutrons, and electrons) is a truly massive factor. Now it's not a vacuum, or is it? It can't be full of air as air is, well—atoms. The more informed reader will also understand the atom itself can be 'split' or—perhaps more accurately put—smashed. And when it is, many, many new particles emerge. Smaller bits of matter? And they, themselves, can be considered as mixes of even smaller particles: among them—leptons, and then quarks, and these quarks are divided into sub-sets which possess different characteristics. We enter here the quantum world; a place defying any simplistic and rational description; a place where only exotic and seemingly 'magical' things happen.

We'll be going into that world too, further along the journey. But for now, I merely wish to thank you for buying my work, for starting a fresh journey with me, and for wanting to explore what I consider to be extraordinary ideas which have an unbelievable consequence for you and I.

Now all I have to do is to show you all this in such a well defined way that you can grasp it too and examine it for yourself after you close the last page. I won't bore you. I won't patronise you. I will offer clear argument and make simple the most mysterious and deeply profound characteristics of our reality. I will refer to works by some of the best minds alive today and the minds of those great thinkers of the past.

I will need to resort to other avenues than what is generally taught and held in reverence to be real and respected science. I should explain why and also apologise right away. When science started out in the western world, curiosity, open-mindedness, and a wonderful spark existed within the psyche of the explorers themselves. I would venture to say that The Royal Society established in the 17th century in England was the first formally recognised group responsible for rationalising scientific research and methods in the west.

Today, over three hundred years later, we live in a world very different from then. We live in an arena, where media domination of what we hear as news is divided between just a handful of companies. We live in a world where even governments bow before the corporate might of businesses and banking organisations. If this didn't affect the way we perceive information, sift it, and retain what works for us, I would not mention it. But it does has an affect on what we learn. Information and knowledge have enormous power and is more valuable than both oil and gold. This is clearly understood by those people who wish to limit the wide distribution of wealth to all. Many similar factors influence the quest of seeking truth and divulging what is found.

All human endeavour involves a set of desires: survival, novelty, reward, avoidance of failure and pain, self-esteem, a future, security etc. These primal and natural desires can stifle the truth to a large extent. Who wishes to speak about spiritual awakenings, initiated through the use of certain psychotic drugs, if they are employed as a chemist working for a large drug company. The expectation within our profit-motive driven world is a type of thinking and perception which favours only materialist views of the things discovered. Anything else is heresy!

In the west, we live in a period where the objects we own, where the income and the houses we live in, the clothes we wear, the cell phones we use, and all the physical things around us seem to form the complete experience of our lives. Life seems mostly about an endeavour to survive better, richer, and to attain more physical stuff to measure our individual successes. Death has become a momentary event swept away into hospitals, and birth itself takes place no longer at home, witnessed and helped by neighbours and friends, but as a clinical event attended to by people 'trained' to do that function to assist you. The human journey through life has been shifted away from a confused, yet often rich and broad experience, into a drive to specialise, fragment, and focus only on limited fragments; human beings have become teachers, or driving instructors, or policemen, or any one of any categorised work roles where harsh limitations are imposed upon their actions and their style of executing tasks. We are now a world of ants, where the ant caste divides us into a myriad of duties all serving a materialistic world, where we are conditioned to recognise that the road to success and less pain, is by amassing money and doing our individual function 'to the book' better or cheaper than someone else.

This 'stage' on which our current generation live, influences not just our lives but also narrows the breadth of our thoughts by limiting our time to ponder the bigger picture. Busy, productive lives can also deny free time to ask and explore the very questions we started asking when we were children, and we will still be asking at the end of our lives as we approach death. I forever ask these questions and I live on the same stage as you. Somehow, I spend time to ponder them properly. I hope to share some of what I have uncovered. In doing this, sadly I have to stamp a black mark over and over again on the dogma within the political discipline of most of the sciences. I will ask you to forgive me a little for that. There is an issue regarding science's powerful capability to discover truth; it has become perverted by a preconceived set of philosophical concepts which limits what it can achieve.

Science today is under-pinned by a philosophical monism called materialism. It wasn't once, but in this century, right now, it is. And because science has provided us with powerful aids and extraordinary insights into our universe, and improved our material wealth and being—we accept it as the new belief system, and unwittingly discard other potentially important research areas which do not immediately fit in with the core materialistic stance.

The cover of this book shows a metal-gagged person for a reason. There are things you think and care about which science dare not talk about. There is a reason for this, and I need to clarify it right away. Perhaps a quickly thought of example will do it: *science will fail to properly investigate reports of people who say they have lived before.*

The 'thinking' which causes this is as follows.

Bodies are material things. You live in a body. You die. As a body is a material thing, it doesn't come back after it decomposes. You were in that body. You don't come back. If science wishes to investigate such reports—as it is already working under a preconceived idea (materialism), it 'knows' you can't come back, and therefore the only likely area of investigation will be into your mind, and why you should think such a 'silly thing' is true or possible. Do you see it? Materialism! Science has no scope or intent to consider if there might be a real truth in what is reported by thousands of people. To be fair, there would be practical problems in such an area of research, but this is not the reason it doesn't happen. The hindrance is caused by something entirely different: *dogma!* Such phenomena is deemed non-phenomena immediately!

I don't know about you, but I still wish to feel a more profound experience of being alive. Sex, TV, movies, good food, and a 50 hour working week seems ever more to be a difficult balance between the pain of what I have to do to survive and prosper, and the rewards I indebt myself to, and pay to myself, as a soother for the mundane, narrow, daily slog to keep myself alive.

I have learnt a long time ago how best to survive here, and many of you know that too. What we sacrifice to do that is we get less and less time to consider why we are here and what happens when we're not. We are conditioned to think we are only here once, and therefore as this is a single materialistic experience, we are taught to do our best to be physically, materialistically, better off. We compete for increasingly limited resources. We subscribe to monetary systems based on thin air. We wish everyone else in the cars in front of us would get out of the way so we can get to where we're going faster.

As a human being, there is one experience you start with and one experience you end with. You cannot remember your birth and it is likely you will not remember your death. The life you live becomes increasingly the life of an ant. No time to think about it, and if you do think about it for too long, the other ants sweep past you seemingly heading for the resources you need too. They don't. They rush towards ultimate death and will have no idea when they get there why they arrived at the end of life so quickly. If they had taken time, even if only by peering through the window of a small book like this in their limited leisure hours, they might have enjoyed the experience more, and gained something other than the certainty of their loss of material wealth. Instead, they might have been enriched by the discovery of clues which point to the fact that their lives were not in vain.

I would like to try and help you, the reader, to discover more to enrich your own lives. I think to do to that properly, I must introduce you to all kinds of thinking and experience which often fails to get into the mainstream. The only accepted and credible means I have to do this, is by referring to our current scientific knowledge, and possibly do it by adding a philosophical view of what is revealed. I must avoid any specific religious perspective as I do not wish to contaminate my use of two objective tools: philosophy and science. If you believe in a prescribed religious teaching and it works for you, I'm not here to talk your religion down. If you believe in nothing, just here now, dead at the end—I do not wish to persuade you. What I wish to do is to reveal a kind of thinking which might make you wonder, and do it in such a way to keep your sense of wonderment open. It is for you to take what you hopefully discover here and fit it into whatever bias or conviction you already have, or simply ignore it.

Ready?

Mol Smith

* * * *

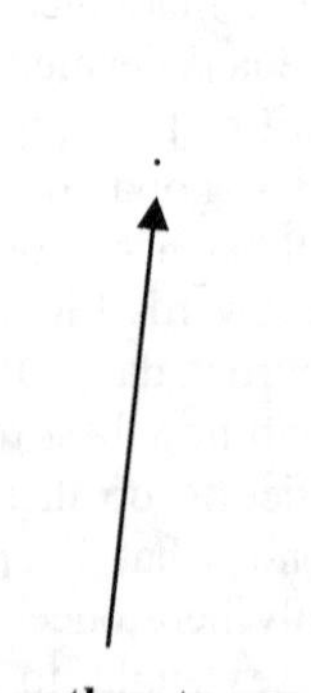

From **then** to now.

Chapter 1: From Then To Now

I suspect most people will already be very much informed of the knowledge we possess about the emergence of life on earth and how we got to where we are today. When I say 'we' I mean people who are informed by science which permeates every aspect of our modern world. But, just in case a few readers are not so informed, or in case some of you wish to consider if I'm kicking my ball around on the same playing field as you are, I'll race through that knowledge in a summarised way in a moment. If you are unfamiliar with any of the terms, perhaps refer to the glossary at the back of the book in the appendices.

But first, a few more words about dogma—a trait which permeates all organised human groups from religion, through policing, politics, business, banking, and the arts. It's a sad fact that we humans are not about truth and the discovering of it for its own sake. We each have a career and our associates to protect. Our individual and collective interests within our peer group often adopt a stance of self-protectionism even if it means the sacrifice of truth. The consequence of this human condition is truth is not really what evolves from long—organised structures and long-established organisations. What emerges is a narrow and distorted part-truth, and the practice of not exploring anything which might threaten in the slightest way the reputation of that organisation.

Dogma is never as strong as in the areas where is it difficult to find real solid evidence for this or that teaching or believed notion. One of the worst disciplines of science for dogmatic practices is in the areas of archaeology and biological history. The reason for this is that the planet does an efficient job of getting rid of the past very quickly and very thoroughly. Some areas of historically-based sciences have theories sold as undisputable truths based on nothing more than a few bones here and a few stones there. Huge reputations are built on massive conjecture, and with evidence which I would say at least—scarce!

I am sixty-four years old. I wonder if any readers out there of a similar age remember the first popular book by a Russian-born American scholar and author, Immanuel Velikovsky? It was called 'Worlds In Collision' and was followed by a second book, 'Earth In Upheaval'. Both books challenged all conventional scientific thinking regarding the history of the earth and the events which shaped it. He was attacked and scorned by the entire scientific community. His works primarily put forward the idea (very simplified here) that the earth has not *slowly* evolved into the world we live in today, but has been shaped and sculptured by catastrophic random events. His books were published in the 1950s. They were controversial. His ideas and concepts flew in the face of established teaching, and possibly were seen to be biased because of his religious belief married to his cultural background. The details of the events he describes, for example—planetary collisions, are still refuted, but the point is not this at all. It's the idea that the Earth was indeed shaped by chaotic and catastrophic random events, which we accept today as true. We merely have evidence now that other factors—not colliding worlds, or religious-associated events caused the modification of Earth. His central revolutionary concept of chaotic and violent influences on the planet was correct even though his evidence for the nature of those events was flawed.

Life on our planet has been all but extinguished several times and has had to restart, more or less, all over again. Super-volcanoes, collision with objects in space, global floods, and ice-ages have been responsible for shaping earth, not some gentle cooling process of a ball of plasma. The universe is not a Swiss watch, and the things which belong in it are subject to chaos and unpredictable events.

I say all this because as I summarise how we human beings arrived here in this century to a point where we can read this book now, we will realise that what is known positions us on very flimsy ground, as I will point out. And, I'm going to explore this concise history along with our understanding of our place in it in a bit more detail throughout the book. That said, for now, here's the story so far. Skip the bits you know.

In the beginning

(A quick start)

Most philosophies, the sciences, and just about everyone on the planet agrees we have not always been here. We understand things start, things end. The start in this case is our reality, which itself is categorised as being this universe—the only one we have any evidence of as existing. Reality equals Universe: our Universe. Conventional thought dictates it wasn't always around and that it came into existence some 14 billion years ago (14,000,000,000 years). It began as a sub-microscopic entity of highly organised, infinitely hot plasma-like material/energy. It didn't stay that way for long. Within microseconds, it underwent massive, inexplicable, rapid expansion. I might add that the initial rapid expansion was followed by continuous inflation ever since, and we are today living on the surface of that expanding entity. We call it a universe.

An important fact to consider is not that this spark came into being somewhere in space and kind of exploded. There was no space. What we call the vacuum of space, itself, is part of the entity (the singularity) which presented itself as an evolving event. It is part of the universe. Science has discovered much evidence to prove this origin, but because we can't create a universe—the beginning of our one remains a non-repeatable event. No matter what the theory and evidence imply, we can never demonstrate this concept is undeniably true, but we can believe it to be so, due to the weight of evidence.

Our universe is considered to be like an expanding bubble with all the stuff we observe within it, from our perspective, existing on the outer surface of the soapy film, including us. This is the reason we detect no centre of the universe. We are not living in a kind of massive explosion where all the bits (the debris) are rushing away from an ignition point. If we were, we could determine a focal point of more densely matter-populated space (the centre) and less densely parts (the distant explosive edge).

This is a very difficult concept for human minds to grasp properly. For a bubble to expand, we imagine space or at least a non-restricting, non-confining

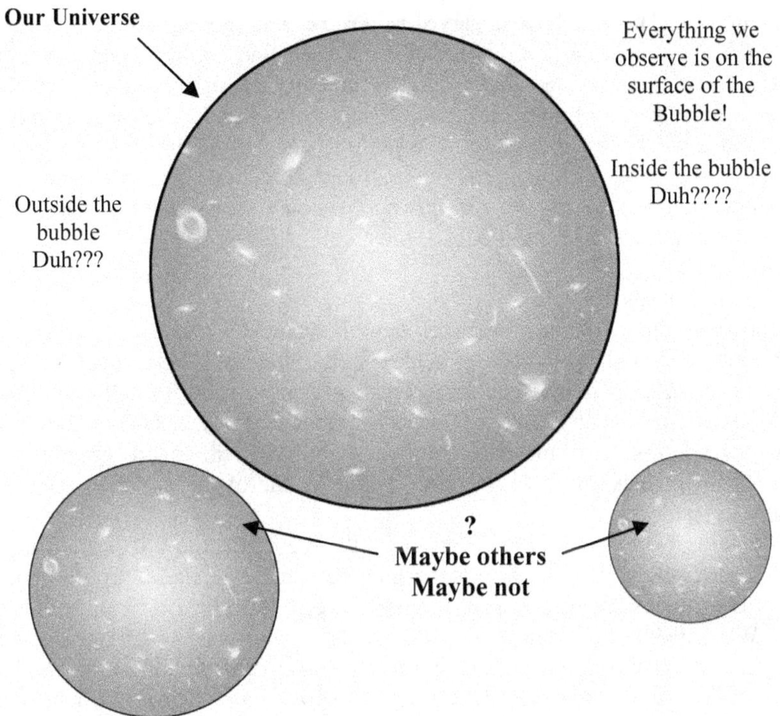

Simplified model—Bubble Universes

something for the bubble to expand and inflate in. As it is impossible to study anything outside our bubble, we cannot realistically consider an 'outside of it'. We may be in an expanding universe where other expanding universes are about to bump up against us, or we may just be the only one expanding in 'who knows what': nobody knows!

It looks about the same everywhere. We don't know if our 'bubble' will continue to expand, slow down, start up again, stop, reverse, collapse or what. We don't know what's driving the inflation.

When we send probes out into space in this direction or that and they 'move' through the vacuum of space, they are confined to the surface of the real entity—the expanding universe, even though to our perception they are moving in a three dimensional physical reality—the one which presents itself to our biological and technological sensors.

Even more astonishing, scientists have an issue about the stuff within our universe, our bubble's surface. It appears we can't see most of it. We can only 'see' the bits which are bright. Most of the bubble, the skin, our universe, is made up of stuff we have completely no information about what-so-ever. It turns out that roughly 68% of the Universe is dark energy. Dark matter makes up about 27%. The rest—everything on Earth, everything ever observed with all of our instruments, all normal matter—adds up to less than 5% of the Universe. What we call galaxies, stars, planets, comets, and ourselves shouldn't be called 'normal' matter at all, since it is such a small fraction of the Universe. The things we observe *are the exception* not the norm!

There are several theories about what so called Dark Matter and Dark Energy are but more data is needed to start to eliminate or prove these theories. I could write a small book just based on exploring what I have summarised so far, and such books exist. But I want only to begin our journey with a commonly agreed starting framework, so I wish to move on. One thing to consider though is the idea of a ***simple*** bubble universe may not be such a good model after all, just like our model of the atom!

Galaxies, stars, planets

Throughout the 14 billion years of our universe's existence, the substance within it (or should I say on the skin of it?), has 'seemingly' organised itself, to become a set of systems. The larger systems we've come to call galaxies—millions of stars centred around massive gravitational centres—are home to planets and less 'hot' material from the 'big-bang'. Stars are gravitational centres for planets. We live on Earth, which is one of these planets spinning around a star—our sun.

Life as we know it exists just on Earth. We don't know if life exists on any other planets. In fact, up to just a few years ago, we had no proof that any other planets existed around any other stars except this one.

We now have evidence they do. When I was born in 1950, the very mention of a belief in other planetary systems being a common thing would be like saying we are visited by UFOs. At 10 years of age, I had no doubt other planets existed. Just like other concepts of the past seemed to start as centric ideas: there is an 'us', our planet, and then everything else spins around that—we repeatedly fall into the trap of being the centre of the ideas we explore, and imagine until proven wrong that in some way, we are the focus of attention. There is no reason for anyone to believe, in the absence of observation or data, any event, process, or entity is a one-off. What we consider as anything 'being' is quite likely *never* to be just a single occurrence. I often find it difficult to understand why religious people should believe in one god, or scientific people should believe there is one universe, or anyone should believe there is one of anything which is a big (profound) thing. Why? If I witnessed the first lightning strike, or the first single ant, would I really think both of these are the only occurrences just because I saw either once only?

This is important to consider. We live in a single universe. We know of no other one. It is unlikely we will ever be able to find hard evidence for another one. It does not mean others do not exist. Given all the other centric ideas which have hampered our knowledge, the idea of a one-off concept of anything seems intuitively wrong.

The conditions for life to exist on earth come down to a statistical value which is equal to winning the lottery about 1 million times in a row. But here we are. I don't believe in beating impossible odds, so already I start to consider the odds are wrong. My next thought is why are they wrong? Maybe it is because the odds are created by too narrow a scientific-only perspective of the problem. The evidence provided for life is only based on life similar to us. Life may exist in conditions different to our own, and be based on a different

In this image of the Hubble Ultra Deep Field, several objects are identified as the faintest, most compact galaxies ever observed in the distant universe. They are so far away that we see them as they looked less than one billion years after the Big Bang. Blazing with the brilliance of millions of stars, each of the newly discovered galaxies is a hundred to a thousand times smaller than our Milky Way Galaxy. The detection required joint observations between Hubble and NASA's Spitzer Space Telescope. The colour version of this image conclusively shows that these are truly young galaxies without an earlier generation of stars. NASA Identifier: SPITZ-ssc2007-15a1

atom than carbon. In fact, when we talk of life, we need to examine quite closely why we consider 'living' organised matter different to other forms of organised matter. Keep that in mind.

The emergence of life

So, this is where we come to the crunch. We are. We think, and because we think, we must consider why that is. Why are we here? And we all know we die. Maybe the whole journey of discovering we live on a planet which orbits a fusion reactor (the sun), which itself spins in a galactic arm of millions of stars, probably around a black hole, is for a profound reason. But maybe the question we are all asking today is not why are we here. Perhaps, our entire quest for knowledge so far has led to the point of no longer asking why we are here, but instead asking how we can continue to survive here—a materialistic aim. To answer this, we need to know clearly what we are part of, and how it evolves due to the consequence of the driving forces influencing it.

How did life get started here?

The quick answer is: we don't know. But because we always start out with centric ideas, the scientific answer is... of course... since we live here—it started here... life! And, again—simply put—it emerged out of the chemistry here, and the lump of that original big-bang stuff that became gravitationally amassed here around our sun. I don't think this is true myself, but let's get on with the dogma first.

Life here on earth is entirely based on a substance we call carbon. All life we know of is based and founded on the properties of carbon. Carbon is an element which really means its atomic structure cannot be reduced to a more fragmented form except of course by sub-dividing it into sub-atomic particles down to quarks, etc.

Carbon is soot. It is also diamond. Carbon is the fourth most abundant element (atom) in the observable soup (5%) of the universe. Most abundant is Hydrogen, then Helium, then Oxygen, then Carbon. I don't wish to give anyone here a chemistry lesson, but if you look at what the carbon atom actually is compared to the three more common elements, you will notice it is the first *solid* element made in abundance after the first event—the big bang. The three more common elements are all gases.

Carbon is produced by stars. It did not exist until stars formed and has come into existence only after of the emergence of other elements (Hydrogen and Helium) and the nuclear reactions taking place in the matter which formed stars. This a huge simplification by me. Maybe take a look at two web references to accept my reduction process:

https://en.wikipedia.org/wiki/Abundance_of_the_chemical_elements
https://en.wikipedia.org/wiki/Carbon

The big thing about carbon (and this is both a scientific concept and a philosophical one) is that it is the first really *versatile element* in the emergence of elements, post big bang. It is capable of combining with itself in a variety of structural patterns. Thus it can be arranged as a pattern of carbon atoms forming the hardest structure we know—diamond, but also the softest and most lubricating substance we know—graphite. So, the first thing stars seem to do well after they produce helium and hydrogen and oxygen, is to produce lots of carbon. I suspect no life exists based on the helium or oxygen atoms due to insufficient possibilities of recombining these atoms in complex ways.

A simple rule in chemistry is an atom will have a maximum number of 8 electrons in any given orbital path or shell. (There are exceptions!) This model of an atom far from represents what an atom actually is. The model itself is part of what I consider to be a major flaw in our attempts to comprehend reality, but it does however allow some kind of basic understanding of why different elements bond together through the process of electron sharing to form molecules and compounds.

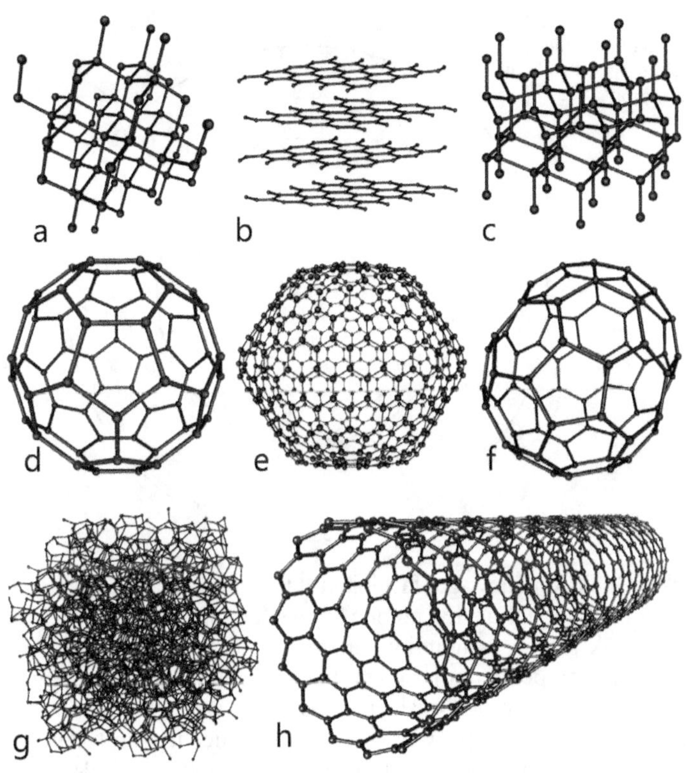

Created by Michael Ströck (mstroeck) on Feb 7, 2006 using iMol for Mac OS X and Photoshop CS2. Released under the GFDL. (Wikipedia).

But we were exploring the emergence of life and the fact it is based on carbon—an atom which is flexible and 'well suited' to re-combining with other elements quite readily. Why carbon? Life is complex. You can't create an organism capable of regulating its internal state, of moving, eating and excreting, and of replicating itself to create offspring, without a wide variety of molecules. You need a central building block which can support more complex branched structures and yet remain strong without sacrificing flexibility to rearrange the whole thing later. Carbon can form four bonds. There are only a few elements that can do this naturally. Oxygen, for instance, will naturally form two bonds (think H2O). The four-bond structure allows a wide variety of possible chains with branches that have branches that have branches. When a bonding slot is unwanted, the carbon-hydrogen bond that usually fills the gap isn't very reactive, so it won't interfere with whatever else might be going on in the area.

Since life can only come into being through combining bits of our universe into sub-structures of it, life can only be built using atoms. More... it is made from the combinations of carbon atoms with other ones from other naturally occurring elements. Life is formed physically from molecules. That is a very profound thing. From big bang, to stars throwing out tons of building

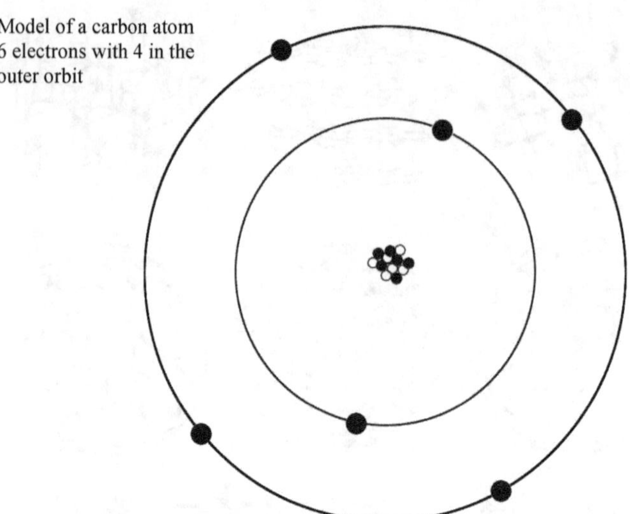

blocks which somehow get fashioned into complete reproducing structures, to us. The issue though is that no one knows for sure how life started on earth. In fact, I think it's an assumption, another centric viewpoint like all the others which have hampered our progress, to even talk about life 'emerging' on earth. Why should it only be considered that it emerged here from some kind of primeval chemical soup? Why not also consider it didn't start here but was *introduced* here, maybe through seeding by comets, asteroids, or stellar dust?

In 1952, an experiment was carried out on how complex organic molecules might have formed under the conditions of early Earth. The early Earth atmosphere is believed to have been composed of methane, ammonia, hydrogen and water vapour. These gases were sealed in an airtight container and then exposed to sparks of electricity to simulate lightning for a week. By the end, a reddish-brown substance had coated the walls of the container. This substance contained 11 of the 20 amino acids used by life on earth. Many scientists now believe the early Earth's atmosphere was composed of carbon dioxide, nitrogen and water vapour. Modern experiments with this mixture of gases produce similar results, but never has a single simple organism formed in any of these experiments. Some scientists believe metabolism, in other words—the ability to break down carbon dioxide in the presence of a catalyst into small organic molecules—was how the first life developed. Other scientists believe the first living organisms were genes--whilst others think it was **RNA.**

The glaring truth though is that even using science's proven tool—that of demonstrating a truth by replicating an experiment that works—fails to produce anything substantial in support of self-emergence of life on Earth.

This failure of science is often focused on by religious people who can then invoke a God or Gods as the instrumental factor introducing life. Science may not have a proof, but it has a conceptual process which seems rational and thus can be shared with many minds to consider its validity. The process of getting from simple organic forms into the more complex ones existing today, including our good selves, is called evolution. Elephants and people did not just come into being as they are now, they developed and formed through many mutations, over millions of years, inside the genetic material (the code) which seems to orchestrate matter into producing living things.

This is not the right time to go into DNA (deoxyribonucleic acid) and genetic coding as most people are aware of how genes are like a computer code made up of 4 distinct chemical compounds (instead of 2 'distinct bits' in computer code): adenine, cytosine, guanine, and thymine. All we need do is accept for now this is part of our perceived wisdom regarding our starting framework of scientific knowledge in respect to reality. We need to look a bit wider yet to cover our historical journey from perceived emergence back before the dinosaurs up until today—our age of computer emergence and their role in our existence.

From amoeba to homo-sapiens

The simplest forms of life we can describe are single-celled organisms—living things, each made of just one cell. They are in the most part microscopic although a few single-celled life-forms can just be seen with the naked eye. Most people will have heard of an amoeba. Most school students will also have encountered a Paramecium {*singular*} (the slipper animal). I've reproduced a photograph of one taken through a microscope below. My very rough diagram beneath it labels some of the key processes which are all part of the one cell.

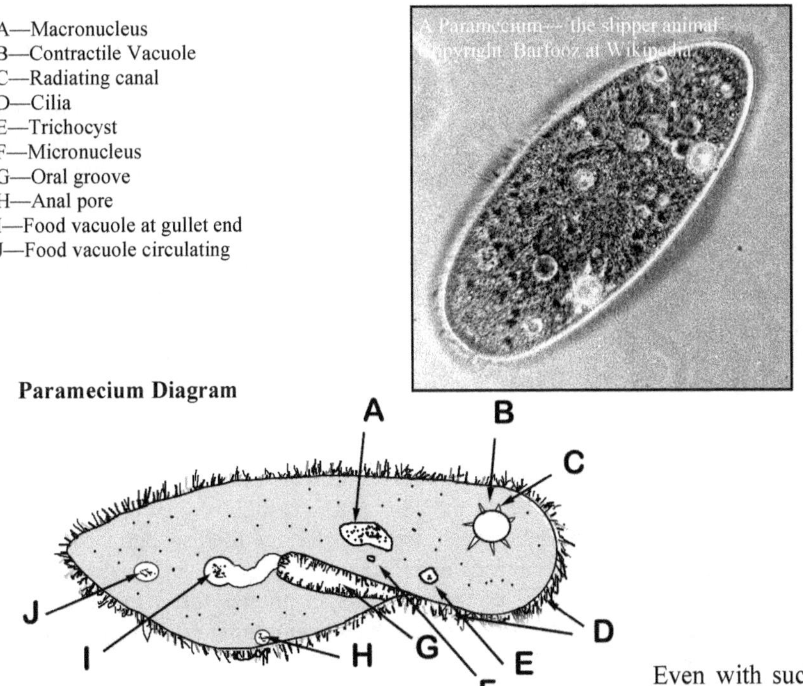

Even with such

a simple organism, it's plain to see very sophisticated mechanisms at work in a cooperative arrangement throughout the cell. A paramecium swims through the water by action of its fine hairs (cilia) arranged along its surface. When the animal encounter an obstacle, it stops, swims back a bit, changes angle, and swims forwards again. This action and how the paramecium senses the obstacle, and how that triggers the stopping and reversing movements, and the changed angle along with swimming forwards again, is something which defies scientific explanation. All the fundamental rules concerning how animals determine external events and sense other entities and how they react to them, are broken. No one knows what mechanism or information exchange system in this tiny creature is at work. What's more is the paramecium is able to remember things too!

Let's not take my word for it, but instead take the word of Roger Penrose, an eminent scientist who was one of the key people working alongside the legendary Stephen Hawking in their younger days:

"If we are to believe that neurons are the only things that control the sophisticated actions of animals, then the humble paramecium presents us with a profound problem. For she swims about her pond with her numerous tiny hair-like legs—the cilia—darting in the direction of bacterial food which she senses using a variety of mechanisms, or retreating at the prospect of danger, ready to swim off in another direction. She can also negotiate obstructions by swimming around them. Moreover, she can apparently even learn from her past experiences—though this most remarkable of her apparent faculties has been disputed by some. How is this achieved by an animal without a single neuron or synapse? Indeed, being but a single cell, and not being a neuron herself, she has no place to accommodate such accessories.

Yet there must indeed be a complicated control system governing the behaviour of a paramecium - or indeed other one-celled animals like amoebas - but it is not a nervous system. The structure responsible is apparently part of what is referred to as the cytoskeleton..."

I have included this little digression into Paramecium, by quoting from his book—'**Shadows Of The Mind**' because science suggests that the brain, made up of many neuron cells, came about through evolutionary processes as a way of organising ever increasing complexity in the living things being evolved. As processes increased in number and sophistication, a kind of central control was required in order to synchronise what was going on within the unit of life. Neurons became the cells of choice within animals to create a messaging, memory, and processing system. We have mimicked this in our own evolving technology using memory chips and central processors, so the emergence of a degree of intelligence being required in living things can be readily understood. There are no neurons in a Paramecium!

The size of the central control systems in animals does not indicate a greater or lesser degree of intelligence. Once again, other factors are involved which are more profound.

So, since even before the dinosaurs, living units had quickly developed rudimentary intelligence to a) control their internal systems and b) to sense and react to external stimuli in more advantaged ways.

An important thing to remember is we have now clearly sited *mind* as part of this physical central nervous network and system. Animals have brains. These are intelligence systems. Minds are our intellects. Therefore, minds must exist in brains. Everyone will agree that brain death seems to leave no sign of intelligence in a remaining vegetative state human-being. We agree it is necessary for a mind to be a part of the animal or a working process in part of an animal for it to be sufficiently aware it exists and functions in our universe—our reality. No mind: no real awareness to negotiate itself in a sea of other organised sets of compounds and elements, and no way to avoid danger to its continuing existence. *Is this true, I wonder?*

One can argue all kinds of cause and effect reasons for why human beings seem to have evolved more intelligent minds than other animals. The opposing thumb, the need to use tools, etc. All or none may have been involved. We can even argue about our intelligence in semantic ways, wondering exactly what is more intelligent about waging war on each other for most of our existence, but this is not fruitful nor helpful in establishing our common starting framework for the rest of this book.

I think we're nearly there.

These are the things we've now established as most commonly thought:

1) The Universe started as a big bang.
2) We live in one.
3) It has lots of galaxies with stars in it.
4) We live on a planet orbiting a star in a galaxy.
5) We are living forms made from stuff in those stars.
6) We have become intelligent.
7) Our intelligence helps us negotiate other structures and events in this universe.
8) Our intelligence is derived from properties of an organ we possess called the brain, which also coordinates our internal workings.

Omissions

I have missed out so much especially with regard to human history and our recollections of it. But it doesn't matter for now. We have our starting point, over-simplified, a bit rushed together, but it's a start. From a common foundation, it's easier to explore the ideas and knowledge we are going to encounter. There is no harm in disagreeing on what comes next, but it is a good idea for each person to consider substantial, evidential and demonstrable reasons why he/she might refute anything said here.

If you are a deeply religious person, you might prefer to consider a creator or God made everything. There is nothing said so far which denies you to continue with that belief. Scientists don't know how we, the universe, and everything came into being. They have personal beliefs. Some believe they do not need to involve an all knowing sentient creator. Others do! I don't wish to get involved with a debate on who is right or wrong. What I have done is create a framework of what is, as far as we know. God, Thor, Odin, Father Christmas, the Universe, itself—thinking about it, or blindly—may have ushered it all in. Who knows? I don't.

What I would like to try and introduce is something completely different—a set of arguments which I believe give room for everyone to consider each and every person living today, and every person who has ever lived before might be more than just flesh and bone. I am going to do this outside of all religious argument and without resorting to bibles and other belief systems. But to do it, I need to pick apart a lot of scientific dogma and I need you to shake yourself free of many accepted notions you might have learnt which are proving to be inaccurate in a quantum-based perspective of reality. Alas, we all learnt science at school based on ideas thought to be accurate and true from people defining the roots of science back in the 1600s, and their followers up to the dawn of the atomic age in the 1900s. We are all still stuck as Newtonians in what has become an Einstein-Hawking-Higgs-Bohr-Planck-Heisenberg-Broglie-Schrödinger-Dirac world. Reality at the small scale is not simple. It is strange. What happens in the tiny spaces of reality also dictates what happens in the bigger spaces—to us!

I believe in knowledge and evidence. It's difficult to find concrete evidence of a god, other than doing that semantically by bringing attention to the extraordinary detail involved in the creation and workings of living forms. That's one way to go, and 'Creationists' put forward such arguments to support their faiths: "Everything is so complex and profound, God must have made it all!"

Science, on the other hand, (but not necessarily scientists) refutes the existence of a creator Science believes life to be a natural occurrence given the universe and the way it works. They view reality as a kind of self-organising whole, where the laws within it govern the incidents and processes we witness, including the emergence of life. They argue there is no need to involve an intelligence in either the creation of the universe or the creation of life.

I'm sure you have your own views. I am not here to subvert them or change them. Whatever you believe, you should be able to find insight and add that to what you already possess from the things I have to say. My perspective is to try and bring information together in a particular way so as to demonstrate a fantastic possibility, an idea more than a wishful hope, something which is often shunned by science, and muted to be a belief that you are asked to take in good faith, and without supporting evidence by religious leaders.

Are our lives simply what they appear to be? Are we born in a cycle of repeated reproduction only to generate new people who then live their lives, reproduce, and die?

Each of us experience a profound and emotional journey in the time we are alive. Much is learnt, not only about the physical world we live in, but how it 'feels' to be a human being in a physical reality shaped by the rules and systems of the universe we inhabit. Yet, it seems, when each of us die—little remains of that journey. Maybe a few memories of you exist in the minds and hearts of surviving family, loved ones, and friends. Some things you did which left material traces, like books authored, or houses built, or photographs taken... these may go on to exist for a few more years. It seems we die and only the fruits of some of our lives go on.

What a waste! How can a reality exist and expend so much energy and effort over 14 billion years to get intelligent life to the point it can witness the fine details of being alive, and then just discard all the important information.

The universe seems to conserve energy. Although it appears abundant, it is treated as a precious thing. And people use energy. Masses of it. Is it all just wasted? (See below).

How Many People Have Ever Lived On Earth? 108 Billion

Year	Population	Births per 1,000	Births Between Benchmarks
50,000 B.C.	2	-	-
8000 B.C.	5,000,000	80	1,137,789,769
1 A.D.	300,000,000	80	46,025,332,354
1200	450,000,000	60	26,591,343,000
1650	500,000,000	60	12,782,002,453
1750	795,000,000	50	3,171,931,513
1850	1,265,000,000	40	4,046,240,009
1900	1,656,000,000	40	2,900,237,856
1950	2,516,000,000	31-38	3,390,198,215
1995	5,760,000,000	31	5,427,305,000
2011	6,987,000,000	23	2,130,327,622

From Population Reference Bureau
http://www.prb.org/Publications/Articles/2002/HowManyPeopleHaveEverLivedonEarth.aspx

The guestimated total expenditure of energy by all humans who ever lived would be equivalent to the work done or energy released (by burning) of 15.120 trillion gallons of oil—about one and half times Saudi Arabia's current reserves in the ground.

Mean Estimate for the weight of living forms, including microbes, currently alive on Earth. Total Biomass = 500 Billion Tons

The idea of death, along with the complete loss of our experiences, stems from a materialistic belief of existing only in our physical bodies, and all of the things learnt and remembered are stored only in the brain. Until the 1920s, this could be considered very evidential. But not anymore! The last 80 years or so of scientific research and the introduction of new theories about reality has cast very large doubts about that. The concept of who we really are needs re-evaluating in light of what is now known. The belief that death causes the loss of all you are, and all you have been, is no longer entirely credible. I wish to demonstrate this fact and to suggest the opposite has greater odds of being true.

My argument begins by questioning our comprehension of reality. Philosophers in the past have thought much about this. Science strives to describe it too. Both approaches encounter a problem, one which makes it extremely difficult to understand reality at all.

* * * *

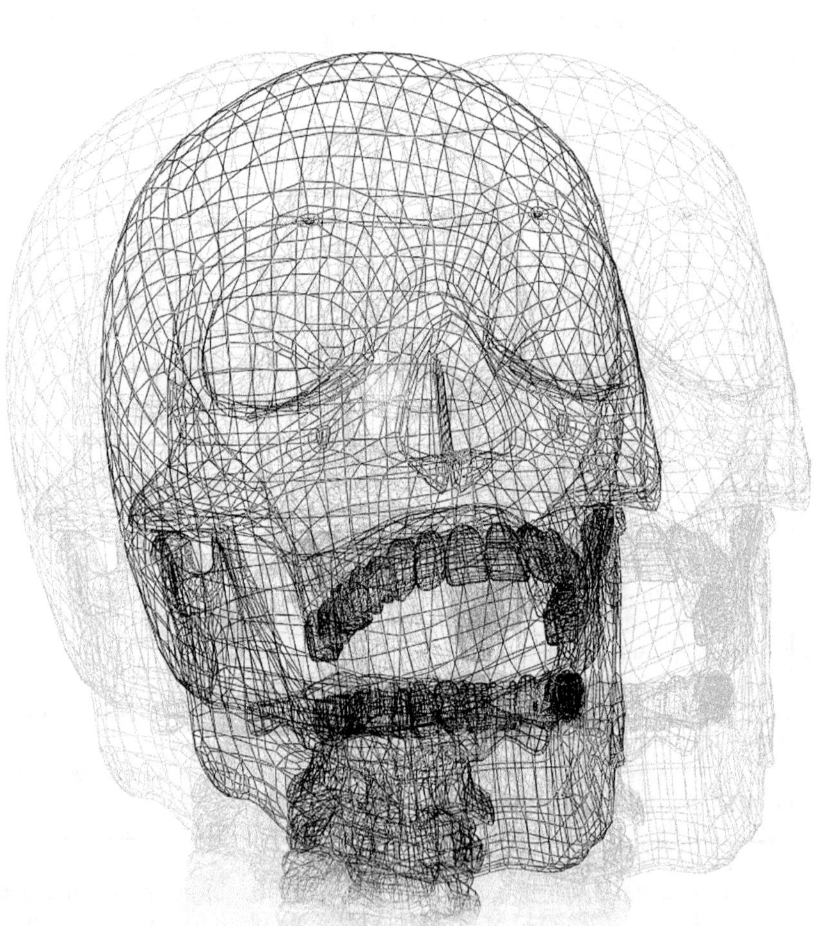

Reality is obscure

Chapter 2: Reality is obscure

Note: *I deliberated about whether to put this chapter before or after Chapter 3. It contains ideals very difficult to understand which has nothing to do with the reader's level of intelligence. I finally decided to introduce the complex ideas here, and then discuss some of them again in the following chapter. Any seemingly repeated information is deliberate and allows me to try and explain or describe the concepts in different ways. If you are confused by what you read, you are not alone. The ideas are very confusing as everyone who knows about them will agree. Hopefully, Chapter 3 will make things a little more clearer after you read this one.*

mol

It is generally accepted when we speak of reality, it means we are describing processes, events, and our existence within our universe. Reality is a concept constructed from a set of notions and reflections beginning as theories and hypotheses which end up being retained or cast off depending on which ideas stand up to repeatable experiments and proofs.

Reality is what most of us can agree it to be using the logic and rationale of our intelligence. Sometimes, there may be unanswered gaps in what we come to agree on, and these holes in our understanding give rise to the argument, debate, and new hypotheses to test out before filling in the gaps. The more we learn, the wider the gaps seem to become though.

But actually, *real* reality has nothing to do with our intelligence. Reality is the way the universe works whether or not we are here to try and understand its working to our advantage or not. Reality exists. *Our reality* is an understanding of it. Our failure to fully understand it so far, equates to only one thing: we do not yet know what reality actually is. Most people are quite happy not fully understanding everything. It only takes a little bit of knowledge to negotiate life from birth to death. Billions of people have lived before us. Most of them never knew they were living in a galaxy, let alone on a planet hurtling around a star. It made little difference to their experience of being alive. A few things helped, I'm sure. Realising there were seasons and understanding when to plant and harvest food is a good example. But it didn't matter if the sun went around the earth, the earth around the sun, or god lifted the sun up every day and put it back in his pocket in the evening. Patterns were realised, noted, and followed. Corn was sown. Food was stored for cold winter months, and people emerged from the less productive periods of winter to welcome the new spring where they would fornicate, breed, and work to keep the show on the road.

Perhaps if no-one became ill or no-one died, or nothing ever happened to cause pain and misery, then no-one would have bothered to ask any further questions about what they are, where they are, and what reality actually is.

And then there is the question of whether or not each of us 'sees' reality the same way. Even given the facts, the same to everyone, we each filter them with beliefs, romantic wishes and subjective fears as we construct internalised models of the part of the universe outside of our minds. We peer out not just through eyes, we listen not only with ears, we also collect the data with a distorted mirror and accept the things which fit selectively onto our internal glass. Reality is a fuzzy common place and each of us holds only fragments of what it really is.

Where once philosophy and religion defined that fuzzy vista, today science and technology redefine it anew. Our computers can map reality in all its exotic forms far more readily than us biological creatures can, but they—the computers—can't understand it even a little. Our human brains grapple with exotic and complex mathematical notions of reality but are out-gunned by our machines. Yet we do understand it (reality) a little. It is because we are sentient. Understanding something has a cause and effect for us. It's important. We need to comprehend it because if we misunderstand an aspect of reality, some tiny detail, could contribute towards something which kills us.

To each of us, reality is a representative thing. It depends on how any given mind accurately builds an internal idea of the outside world; the more accurate the internal model, the more advantaged the person with the better model has of surviving.

The new reality

One of the most critical experiments (demonstrations?) a person can witness, immediately showing how profound the 'bits' of our universe actually are, is the double slit experiment.

If you look at the three diagrams on the next page, I'll explain how a simple experiment reveals, in an easy-to-grasp way, how matter is not what we originally thought it to be. In the first diagram, we fire marbles at a screen with a slit in it. Some of the marbles bounce back, while other pass through the slit and strike the target, leaving marks where they impact. If we put a second slit in the screen, a second pattern of hits is recorded.

If we now fire coherent light, say—from a laser, at the screen with a single slit (Diagram 2), the leading edge of the light wave leaves a bright mark on the screen, with gradual falling off of brightness either side. If we fire the light through a screen with two slits in it, the two light waves, one passing through one slit, and the other passing through the other slit, interfere with each other causing an interference pattern to be created on the target.

Now, here's the start of the biggie. If we fire electrons, bits of atoms, at a screen with one slit in it, they behave just as the marbles did and leave an impact line on the target. (Diagram 3). But, if we fire electrons through two slits, we get a surprise. They do not leave a pattern on the target like the marbles did. Instead they leave a pattern on the target similar to the one where light waves were fired at the screen.

If electrons are solid bits of atoms, we would expect to see two lines on the target screen, just like we do with the marbles. When we have just one slit, the pattern for electrons is indeed, just like the marble pattern for one slit. But the fact, that two slits produce a pattern the same as when we fire light waves, means electrons appear to have the ability to behave as solid particles and as

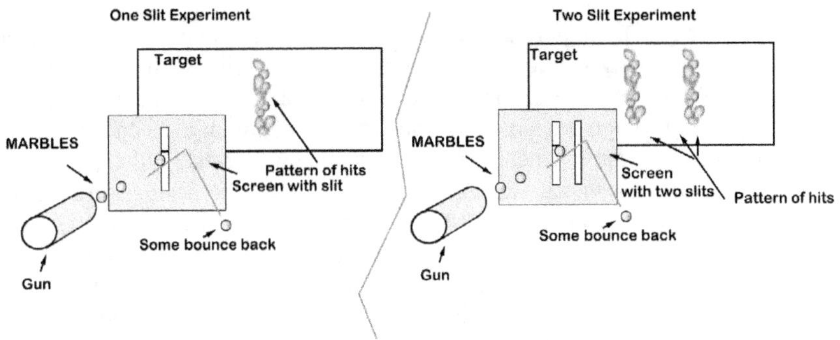

Diagram 1.

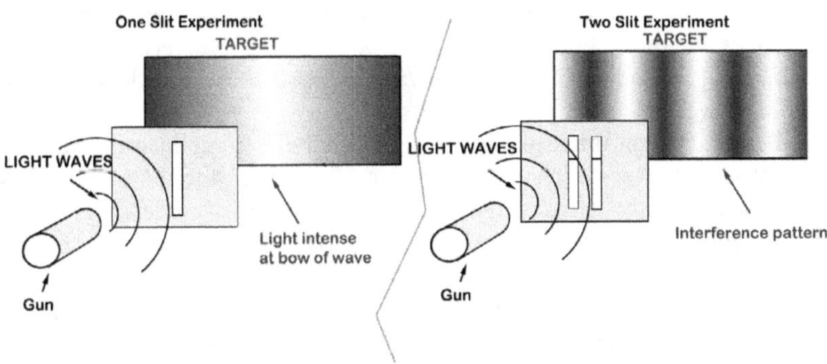

Diagram 2.

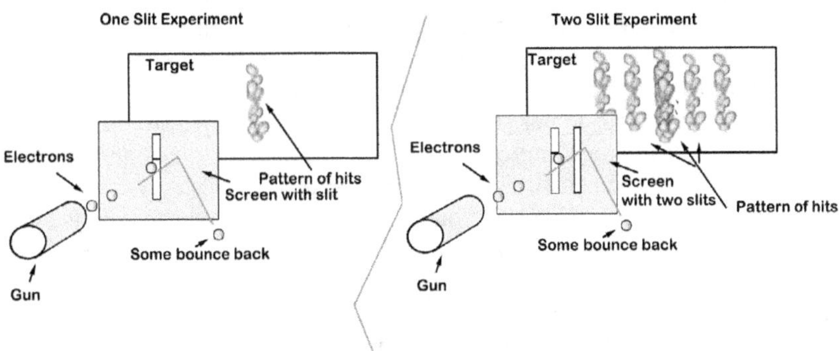

Diagram 3.

waves of energy. The important fact to understand here is ***one electron is not interfering with another one;*** it's more like it's interfering with itself! This is proven by just firing one electron or one photon at any given time, so only one electron passes through the slit/slits in any given fraction of a second. It seems when one electron approaches the slits, it somehow 'realises' there are two possible paths, turns into a wave, and passes through both slits—producing two waves which interfere with each other (creating the pattern on the target), before becoming a particle again and hitting the target in a single position.

If this isn't strange enough, *mathematically*—it works out the particle might actually travel through one slit, both slits, the other slit, or none of them.

This is baffling. It was decided to really try and find out through practical experiments exactly what is going on. A detector was set up to monitor which slit the electron took to pass through the screen. But something very odd happened then. The electrons stopped behaving as though they were light waves and went back to behaving as though they were only particles. No inference pattern was left on the target, just two impact lines representing particle hits on the screen. The electrons behaved exactly like the marbles go through either one of the two slits. *The mere act of trying to detect or 'observe' which slit the electron passed through has influenced the behaviour of the electron!*

At first, you might think the detector, itself an electronic instrument, may have emitted a wave which interacted or 'informed' the particle of its existence. Scientists thought that too, so in 2002, a group of researchers set up the experiment in a way the electron could not possibly receive information about the existence of an observing instrument. The setup was on a much smaller scale: a single photon was emitted, and an interferometer to observe the wave-or-particle behaviour is either inserted or not inserted.

The insertion of the interferometer took only 40 nanoseconds (ns) while it would take 160 ns for any information about the configuration to travel from the interferometer to reach the photon before it entered the slits. This means in order for the photon to "know" if it was being watched, information would have to travel at 4 times the speed of light, which is impossible (the speed of light is the universal speed limit). The electrons behaved as they did before: weirdly! This proved the electronic detector was not 'informing' the particle.

This one experiment illuminates the counter-intuitive nature of matter and its odd behaviour at sub microscopic scales. The problem is we don't observe such strange behaviour in our macro world (the Classical World), but since we are made of those atoms and those electrons, a state of quantum activity is somehow likely to be actually going on in you and me! We may not be aware of it, but somewhere, everywhere where matter exists, in your foot, in the table over there, in your brain, that star—the underlying rules, processes and mechanics of all matter's behaviour is also being influenced by exotic rules and interactions which defy rational thought.

It isn't just the double slit experiment which demonstrates this. Another aspect of matter at the quantum level arises, which is not just theoretical but a proven and mind-defying thing. It's important now to talk about particles, atoms and quantum states. I think both myself and the average reader will not really appreciate any mathematical explanation, so I'll do my best to cover it with analogies and mind-models.

Quantum entanglement

There are aspects regarding our universe (our reality) which are deemed absolute truths. Maybe some future concept or experiment will prove even these *absolute* truths are in fact wrong, but so far they hold up under the most challenging scrutiny and practical application. The most essential one concerns the speed of light. The truth is this: nothing can move faster than the speed of light. Even if you build a spaceship which could achieve the speed of light and started to pour more energy into its drive to make it exceed the limit, it will go no faster. What would happen is that all the extra energy you put into the spacecraft's engine would simply increase the mass of the ship itself. No, it won't grow bigger. It just gets well... heavier! That said, we use light to carry information. As does the Universe. The reason you see the sun or that distant star is because light is carrying information about it to your eyes. Since light cannot exceed a certain speed, information transference cannot either.

Photons move at the speed of light. A photon is an elementary particle, the quantum of light and all other forms of electromagnetic radiation, and the force carrier for the electromagnetic force, even when static via virtual photons. The effects of this force are easily observable at both the microscopic and macroscopic level because the photon has zero rest mass; this allows long distance interactions. Like all elementary particles, photons are currently best explained by quantum mechanics and they exhibit wave–particle duality, exhibiting properties of both waves and particles.

Photons can be absorbed by nuclei, atoms or molecules, provoking transitions between their energy levels. The absorption of photons can even break chemical bonds, as in the photo-dissociation of chlorine; this is the subject of photochemistry. Just like other sub atomic particles, photons have measurable characteristics. We need to consider a few of these: spin, axis, direction of motion, and polarisation.

A photon's axis and its direction of motion are directly linked, making it impossible to change one without changing the other, much like a gyroscope. In all cases, a photon's axis must be 90 degrees to its direction of motion. Since photons travel at light speed, their spin is limited to two states: clockwise or counter-clockwise. These states correspond to left-handed and right-handed photons.

Polarization is the direction of photon oscillation. In physics, 'oscillate' means to vary between alternate extremes, often within a set time limit. For a photon, its polarization is the orientation of its axis, which in turn is linked to its direction of motion.

When a photon, say from polarized laser light, passes through matter, it will be absorbed by an electron—increasing and exciting the electron's energy state. Eventually, and spontaneously, the electron will return to its ground state by emitting the photon. By firing photons into certain crystal structures, we can increase the likelihood of the photon not emerging as a single unit but instead it will split into two photons, both of them with longer wavelengths than the original. A longer wavelength means a lower frequency, and thus less quantum of energy. The total energy of the two photons must equal the energy of the original photon fired from the laser to obey the rule of conservation of energy. When the original photon splits into two photons, the resulting photon pair is considered entangled. Normally the photons exit the crystal such that one is aligned in a horizontally polarized light cone, and the other is aligned vertically. And although they exist in reality as discrete real particles and may be divided by an infinite distance, if you measure or affect one of the entangled photons, it will have an immediate instantaneous effect on its entangled partner. Although time and space separate the photons, somehow, information 'appears' to pass from one to the other to ensure they remain in opposite states and thus obey the rule of conservation of the energy which gave rise to them.

Various theories have been put forward about how this information exchange can take place faster than the speed of light. One theory suggested that when the photon splits into two, each new photon leaves with hidden variables associated with it, such that either one does not need to know anything about its entangled one to maintain itself in an opposite state to the entangled one. Unfortunately, this theory has failed all proofs and is thereby considered an unlikely explanation!

Although it is paradoxical, entangled systems like photons have a way of influencing information and characteristics in the entangled partner faster than the speed of light—so fast in fact, that no speed can be measured: it is exactly, precisely, instant—irrespective of whether the distance between the two systems are a nanometre or a billion light years*! (Note: very large separation distances have not yet been proven by physical experiments).*

This effect is of major importance to many areas of commerce and industry and is therefore receiving a lot of attention and funding. It offers concepts and possibilities of super-speed computing, and faster-than-light long distance communication. Recent experiments have therefore gone even further.

The following is from Wikipedia and a follow up of its references:

In a 2012 experiment, "delayed-choice entanglement swapping" was used to decide whether two particles were entangled or not after they had already been measured.

In a 2013 experiment, entanglement swapping has been used to create entanglement between photons that never coexisted in time, thus demonstrating that "the non-locality of quantum mechanics, as manifested by entanglement, does not apply only to particles with space-like separation, but also to particles with time-like (temporal) separation". ***My note: information can travel through time!?***

In August 2014, researcher Gabriela Barreto Lemos and team were able to "take pictures" of objects using photons that have never interacted with the subjects, but were entangled with photons which did interact with such objects. Lemos, from the University of Vienna, is confident this new quantum imaging technique could find application where low light imaging is imperative, in fields like biological or medical imaging.

From Nature Magazine 27 August 2014:

Dr. Anton Zeilinger and his colleagues based the technique on an idea first outlined in 1991, in which there are two paths down which a photon can travel. Each contains a crystal that turns the particle into a pair of entangled photons. But only one path contains the object to be imaged.

According to the laws of quantum physics, if no one detects which path a photon took, the particle effectively has taken both routes, and a photon pair is created in each path at once, says Gabriela Barreto Lemos, a physicist at Austrian Academy of Sciences and a co-author on the latest paper.

In the first path, one photon in the pair passes through the object to be imaged, and the other does not. The photon which passed through the object is then recombined with its other 'possible self'— the one that travelled down the second path and not through the object—and is thrown away. The remaining photon from the second path is also reunited with itself from the first path and directed towards a camera, where it is used to build the image, despite having never interacted with the object.

Love it or hate it

Einstein found it difficult to accept several aspects of the breaking-ground concepts of Quantum physics. It is not difficult to see why. The most successful theory of all science in all of human history is constantly bringing about startling innovation and new technologies, such is the power of practical application of the effects we find in the quantum world. And yet no-one, and I mean no-one, no scientist, no theoretical particle physicist, even God himself can get their heads around what we have discovered as 'true' aspects of our reality. They defy understanding by our 3D minds, where all our senses lead to the building of a 3D internalised shadow model of a reality which in actuality has facets and aspects which fail to 'compute' there-in.

Sadly, our minds are not able to understand reality even when we manage to pierce the veil with our intellect and glimpse powerful traits and processes existing just beyond our reasoning capability.

Some might say, don't worry about understanding it. It works. Use it. And we do. But if you want to understand what you are, and if you wish to consider whether your life is a complete waste of time, because you die, and everything you experience dies with you; if you wish to understand how so much of our thinking about the true nature of ourselves is still stuck in that old 1600s Newtonian world—then you have to look further than Newton did, and start to think the way Quantum States think (if you will excuse my naughty play on words).

There is an implication here about quantum entanglement which is often missed. It's demonstrated through the double-slit experiment. The fact that the travelling electron seems to behave as a definite particle when it is measured or observed, instead of exhibiting wave behaviour, suggests the observer and the particle itself are in some fashion 'entangled'. It is *not* the measuring instrument/detector and the particle in some way interconnected; it is the fact of a mind, in this case the mind which set up the observation to make a determination of the event, or the mind which will ultimately register the outcome—it is this which is influencing the outcome and the behaviour exhibited by the particle: *the observer and the observed are caught within a system as a whole.* The presence of a human intellect is, in an inexplicable way, having an affect upon a piece of the universe through thought alone. Or... the inverse is true! This is very profound. I am not suggesting some kind of mind over matter conscious control, like moving objects around on a table (telekinesis): I am saying the existence of mind is already entangled to a far greater degree than imaginable with the reality we are involved with! The human mind seems to be, by its very presence, able to remove 'multiple possibilities' from sub-atomic particles and endow them with certainty! But, maybe something else is going on instead.

Quantum theory was dreamed up to describe the strange behaviour of particles like atoms and electrons. For nearly a century, physicists have explained the peculiarities of their quantum properties such as wave-particle duality and indeterminism by invoking an entity called the wave function, which exists in a superposition of all possible states at once right up until someone observes it—at which point it is said to "collapse" into a single state. Few people have been comfortable with this notion simply because the 'wave-function' is merely a mathematical abstract—a dreamed up way of working on quantum system problems and predictions. No one actually believes the wave function is a real attribute of the physical world.

In the 1950s, a new theory came along, pioneered by theorist Hugh Everett. It became known as the many worlds theory. His theory *does* treat the wave function as a real physical property—a fundamental characteristic of the real world. In this theory, the universe splits into pairs of parallel universes every time a wave function collapses. The problem with this theory, which is often cited, is the fact that the Universe must expend an immense amount of energy or 'effort' constantly creating parallel universes on the fly. Whenever we consider properties of our physical reality, the universe seems to be quite conservative about energy use, so the idea might have flaws in it.

Now, recently, a brand new interpretation of the many worlds theory has been put forward by Lev Vaidman at Tel Aviv University in Israel. He proposes there already exists multiple almost identical universes (probably infinitely) that were all created at the time of the big bang. The idea is all these universes occupy the same physical space, and interact with each other via repulsive forces between corresponding particles as soon as they diverge a tiny bit.

The approach has some advantages over the standard many-worlds interpretation, says Lev Vaidman, who has worked extensively on the Everettian approach. For example, the many-worlds view struggles to explain probability in a world where everything that could possibly happen does happen. With many interacting worlds, probability falls easily out of the mathematics.

In the "many interacting worlds" theory, nearly identical particles in parallel universes bump into each other to create all the weird quantum effects we observe in our world. Howard Wiseman at Griffith University in Australia and colleagues demonstrated mathematically that the theory predicts the results of the classic double-slit experiment. With many interacting worlds, each photon exerts a force on all the others, nudging the photons in our world on a specific, slightly different trajectory. Wiseman and colleagues showed with just 41 worlds, you can produce the same pattern as in the experiment.

What scientists miss, philosophers probe! So the profound question here is why would an infinite number of near identical universes all be created in the first place? It's quite hard to explain any reason for the creation of one universe; an infinite number is... um... infinitely more difficult. Or so you might think. I personally think this theory, if it can be proven true, opens a chink in the curtain which hides ultimate truth from us.

If you wish to win at roulette, put a chip on every number. You'll definitely pick a winner, but of course you'll need to spend a lot more than you win to gain that outcome. Maybe the universe is doing something similar? With multiple universe comes the possibility where everything that can happen, will happen... if not in one—then another. It strikes me right from the outset, our universe/multiverse might seek a 'desired' outcome... or... and this is as important—it might seek ultimate possible novelty.

Who doesn't hate boredom. It's a real killer. If you can create something, yourself, for example, why not do it multiple times and capitalise on the widest possible range of experiences possible. Maybe, from a philosopher's point of view, the reason you are here—the reason reality is here is simply that: *to witness infinite novelty rather than nothing!*

And perhaps my main point really is, with perhaps with multiple universes also comes multiple versions of you. In some, you may already be dead. In others, maybe you were never born. In one, technology might have advanced humankind more than corporate businesses are exploiting it, and defined a way how you can live forever?

The bottom line is weird quantum behaviour, evidentially, suggests more possibilities and theories that we live in some kind of infinite possibility multi -universe type reality. This sounds more acceptable than a single universe where a fudge-factor mathematical notation somehow, almost impossibly, ends up being a real trait of sub-atomic systems. I'll summarise the importance here. Of all the theories and all the knowledge that science has discovered, right now, its biggest unstated claim is: even if you drop dead here right now, somewhere you are still living. Where? In another universe right on top of you.

It may even be possible once these theories are further developed and tested to realise communication is possible between one universe and another one—if not all of them. I'm not suggesting you can contact yourself in another one. I suggest experiments at the quantum level could prove interaction *is* taking place between sub-atomic matter in each universe.

A further conjecture I would like to suggest is that you are not one entity but a multi-entity, a composite of all experiences in all universes. Your journey in this reality is to experience all that you can be in a physical realm within its constraints.

Of course, in our everyday classical world, these exotic ideas seem far away and might be considered simple artefacts of wild and romantic imaginations. But I'd like you to consider this. Imagine popping back two thousand years and explaining the following to the Romans:

You: Yup, Mr Roman. It's true. You sir, are a living form sitting on a spherical mass spinning on its axis at 1,040 miles/hour and whizzing around a massive fire ball 875,000 miles across 93 million miles away. What's more, the planet you're on and the sun are part of a cluster of similar such systems, a galaxy so vast it would take you 100,000 years to cross it travelling at 186,000 miles per second. And, Mr. Roman, that galaxy is probably spinning around a black hole so massive it will eventually swallow all the millions of stars in our galaxy.

Roman: I don't understand.

You: And that galaxy is just one of millions of such systems all flying away from each other at up to 186,000 miles per second.

Roman: I don't understand.

To the Roman, such ideas appear preposterous and impossible to grasp. And when you read it aloud, on the face of it, even to me—it seems well... just daft! But that's what science has constructed to be an observational truth which can be demonstrated to be... well...true! Let's consider our Roman is a very bright dude. It might go something like this instead:

You: And that galaxy is just one of millions of such systems all flying away from each other at up to 186,000 miles per second.

Roman: Crikey. Yes. I see it.

You: You get it?

Roman: Yup. But...

You: Yes..?

Roman: Why?

You: Why what?

Roman: *Why is it this way?*

And here's the rub. Science is not here to ask why. It's here to ask what and how. Science will never be tasked to answer why. It's a philosophical question. And this is the big failure of science, not accepting it's not enough to discover just what, and thinking it's a complete discipline and tool to uncover truth. To know why something is happening is more important than knowing what is happening. If you woke up in bed during WWII with bombs dropping down onto the street outside, you might be able to take action to avoid being blown out of existence. But if you knew why bombs were dropping outside, you might be able to stop it. Or, if you knew why bombs are going to be dropped on the street outside, you might be able to prevent it.

Science can not know why there is a universe or multiverse... ever! Not unless it changes its dogmatic approach and accepts a less-materialist perspective instead of its existing and classic solo-materialistic one—an approach which forever limits its power as a knowledge based tool.

* * * *

Many worlds—multiple realities

Chapter 3: Many worlds—multiple realities

A kind of strangeness exists between curious and probing human observers and the sub-microscopic world they probe—a strangeness not explained by any conventional aspect of science usually applied to the macro world. Nor can the strangeness be swept away due to erroneous data or faulty equipment. The implications are not made any clearer by the exaggerated novelisation of science fact by imaginative people extending accurate observation into the creation of compelling stories to seize popular public attention through media headlines and fantasy movies.

The point is: *is there really a connection between sub atomic matter and a sentient mind or are we misunderstanding the conclusions?*

Something isn't quite right. A simple string of logic does not fit in with conventional understanding. Why should an electron or a photon seem aware a human mind—nay, an intellect, a sentient observer and a participant in reality—is trying to determine the outcome of an event which should be... well... somehow undetermined, uncertain, and its outcome part random. It is not a case of the outcome being determined by a flaw in the experiment or measuring system, or by conscious attention to fine detail. The enigma is akin to an observer's presence affecting the result of a spinning roulette wheel and the uncertainty of the ball finally resting in a red or black slot in the wheel. In the double slit experiment, the observer's presence of the electron selecting a slit causes the electron not to play the game of uncertainty at all. In our analogy, the roulette ball always finds black, or—to put the illogical result more aptly— it always finds both red and black with the one ball!

It is a problem noted and written about for decades, and it won't go away. Ultimately, humankind is left with only various conjectures about this very tiny event which has such a big impact on our comprehension of reality, whilst the scientific community and students entering it, are left in a position of having to move past it because the mystery is not answerable. They know a strangeness happens here, and many theories exist to suggest why, but all of them have consequences appearing as absurd, and are thus unpalatable to accept.

We chip away at the edges of the problem and avoid the central issue, hoping as we go, what we discover may satisfy collective rationale in a way we can feel comfortable about our definition of reality, and confident about the work put into a quest to understand it over the last 400 years. In the end, in real application and the return of investment into the scientific journey, it doesn't matter about why something works. It matters more about the fact something does work in the way described. By this, I mean it achieves a particular outcome so things can be made by exploiting known effects rather than fully understanding them. We can manufacture tunnelling-effect diodes, LEDs, and many other devices benefitting our material existence, even though a complete understanding of the reactions taking place inside each, eludes us.

Everything we've come to produce physically in this new 21st century era exploiting quantum effects can be traced back to just one original observation—the double slit experiment, and the enigma of intellectual presence having an impact upon the way it refuses to reveal an understandable core truth about reality.

I am not the only person to realise it doesn't all add up. Everyone who looks at the issue closely, knows it too. Many scientists can't voice an opinion. Most students are obliged to adopt theories which are no different from the improvable belief systems we had to suffer in the past. I believe science itself stands at the edge of corruption and loss of purpose through swapping the quest of comprehension for the ambition of commercial application.

And the problem grows deeper the more you focus on this core issue. It affects not only our understanding of what atoms are, and what the nature of matter really is, it undermines the entire framework of science as a tool to describe reality as it actually is. Science, as a processing tool, is strong at revealing how to manipulate matter in beneficial ways, but it is now in danger of displaying mounting weaknesses because it's unable to break free of some of its constraints. Science needs to be updated to adopt the examination of ideas and concepts which are the keystones of other more human-orientated domains, and ways of thinking.

Wave Function

(wiki-complex definition): https://en.wikipedia.org/wiki/Wave_function

I believe this is something everyone should try to understand. For anyone unfamiliar with quantum physics (most of us ordinary human beings are actually in this group), I need to explain why you often hear this term—wave function.

A wave function in quantum physics describes the quantum state (mixed up state) of a system of one or more particles. The mathematical function contains all the information about the system considered in isolation. Quantities, which are normally measured, such as the average momentum of a particle, are derived from the wave function through mathematical operations describing its interaction with observational devices.

The wave function behaves qualitatively like other waves: water waves or waves along a vibrating string. The function is, after-all, a wave-related equation. This explains the name "wave function". It is *often mixed up* with the notion of wave–particle duality. There is *no real wave of the measured system in physical space*; it is a wave in an abstract "mathematical space", which can be represented as 'configuration space', or can be represented as 'momentum space', and in this respect it differs fundamentally from water waves or waves along a string.

There are two prominent interpretations of the wave function dating back to its origins in the 1920s. In one view, the wave function corresponds to an element of reality objectively existing whether or not an observer is measuring it. In an alternative view, the wave function does not represent reality but instead represents an observer's subjective state of knowledge about some underlying reality.

This takes a lot of thinking about.

The idea is an unmeasured atomic or sub-atomic particle is able to occupy many states and spatial positions simultaneously. When we say a particle of the quantum world occupies many states at once, what's really being referred to is the element's wave characteristic (instead of 'wave', think 'field'—it's a better description). A wave function can be viewed as a space occupied simultaneously by many different possibilities or degrees of freedom (abstractly). *It is a mathematical device*—an ultra sophisticated slide-rule. **It is NOT a real thing but a computational method!**

If a particle could be in position (x, y, z) in three-dimensional space, there are probabilities it could specifically be at (x1, y1, z1) or (x2, y2, z2) and so forth, and this is represented in the wave function, which is all of these possibilities added together.

Consequently, even what we'd normally (deterministically) consider as empty space, can now be examined mathematically by assigning to it a wave function and, as such, the solution to the equation will contain some possibilities of it (the empty space) not being empty at all. Sometimes this manifests real "virtual" particles. Also, a bunch of particles can share these states at the same time, effectively becoming instances of the same particle entanglement... er... that is 'on paper'.

You have to grasp this: the math model, the wave-function, the computation itself, conjures up *virtual particles* in empty space. "Wow. What—like banks create virtual money from debt?"

It's possible to strip away all of this indeterminateness. To do so is actually quite easy. Wave functions are very fragile. They are subject to a "collapse" in which all of those possibilities become just a single particle at a single point at a single time... er... mathematically. This is what is used to represent 'notionally' what happens when a macroscopic human-being attempts to measure a quantum mechanical system: the abstract wave drops away and all that's left is a well-defined thing.

It's important to note, there is no evidence or suggestion by anyone there is a real physical wave around the system or particle being measured. It's more that the things in an atomic or sub-atomic world are influenced by the desire and attempt to *determine* them, and they deny human certainty of say knowing an electron's vector as well as its position in space. It is as if until we try, they are quite happy existing as phantoms, evading manifestation of being absolute and precisely 'anything' at any exact point in space, until a mind tries to determine that thing, and then only a few characteristics can be teased out as being probable, but not certain. And this can only be attempted by using a mathematical formulae with a made-up factor in it called *Psi.*

The actual behaviour of any individual photon is totally random and unpredictable, not just in principle but even in practice. Although the tossing of a coin, for example, is random in practice, if we knew precisely everything about the force, angle, shape, air currents, and all the factors influencing it, we could, in principle, predict the outcome precisely. The behaviour of a sub-atomic particle, however, is random on a whole different level, and can never be accurately predicted. This is unacceptable to science. It denies accurate exploitation of the effect. A work-around is used: *Psi.*

Thus, in the quantum world, it is not possible to predict a single definite result for an observation, only a number of different possible outcomes, each with a particular likelihood or probability. Physics has therefore changed overnight from a study of absolute certainty, to one of merely predicting the odds!

The reason we do not see the effects of this on a macro scale is because everyday objects are composed of billions or trillions of atomic particles. Although the position of each individual particle may be highly uncertain, because there are so many of them acting in unison in an everyday object, the combined probabilities add up to what is, to all intents and purposes, a certainty in a macro reality.

In order to reconcile the wave-like and particle-like behaviour of light, its wave-like aspect needs to be able to "inform" its particle-like aspect about how to behave, and vice versa. It was the Austrian physicist Erwin Schrödinger, along with the German Max Born, who first realized this and worked out the mechanism for this information transference in the 1920s by imagining an abstract mathematical wave called a probability wave (or wave function), which could inform a particle of what to do in different situations. Schrödinger proposed a ground-breaking wave equation, analogous to the known equations for other wave motions in nature, to describe such a wave. Born further demonstrated the probability of finding a particle at any point (its 'probability density') was related to the square of the height of the probability wave at that point.

What does all this tell us about reality?

It tells us we live in a universe where things have seemingly not been predetermined. All possibilities, all configurations of possible realities and futures are not set but are 'up for grabs'. There is no fate. The universe itself appears blind to its ultimate outcome.

The Measurement Problem

To visit once more the issue of an observer trying to determine measurements of a particle or anything in the atomic/sub-atomic microscopic realm, we need to fully understand the concept of superposition. Think of a superimposed photograph, a sheet of photographic film that has been exposed several times so you have many pictures all overlapping each other. This is an analogy for superposition in quantum mechanics. If you think about an elementary particle —a photon, an electron, possibly inside a container where you've trapped it, the particle could be at any number of positions in the device you trapped it in.

We know the mathematical tool of applying a wave function can be exploited and it will describe all of the positions the electron can be at, but it cannot yield information to suggest any one of those positions is more or less likely than any other position. They have an equal reality; that is, before you actually measure the electron, it could show up in any of the places the wave function predicts it to be. When you measure it though—when you interact with it, when you observe it—it settles in one definite position. We see only this one position. However, there's only a probability it will have a certain position, so if you measure it again in exactly the same way, you'll get a different position for it and thus—a different measurement result.

At the risk of becoming boring and repeating myself, you need to understand every person on earth is baffled when trying to imagine the quantum world of microscopic particles. No one can actually comprehend it. However, it does not prevent us from using the observed effects. Young kids press buttons on their TV remotes and work the television. Most of them have no idea of how the remote control actually does that in any detail. Well, that's the way it is for we humans and quantum mechanics.

The wave function is used to predict possible outcomes. It is a device which works, but is it accurately telling us anything substantial about the nature of reality? For any atomic particle or system extending from the almost phantom world to have a cause and effect in our macro world, we speak of wave function collapse. This is only a way of trying to explain how particles, floating around in non-determined positions in space, end up a within a solid, physical, 3D macro object. So, according to science: the graphite tip sitting at the end of my pencil is there only because sub-atomic particles, previously existing in positions of space unknown, are now considered to be there because their abstract wave functions collapsed.

Clear?

No? I thought not.

Nor, I believe should it be. The reason you think it looks like fog, is because it is!

Scientists have fudged something to get around a big problem, one saying... 'er... we don't know, but if we use this bit of maths, we can exploit the effects witnessed'.

In fact, the wave function has been a cop-out for such a long time, many people think it is part of the quantum world itself. They believe a kind of wave exists for any sub-atomic particle, and that its wave can collapse to render it as a particle in real physical space, at its measured (detected) position. Bunkum! No-one in the world with any real insight will state the wave function is real. It neither explains, nor models through any kind of human insight, the piece of reality it's applied to when providing possible predictions about outcomes.

Somehow, the maths used in the classical world, the wave equations originally applied in computations concerning fluid flow, ocean waves, sound-waves and the like, could—with a little adjustment—be used in quantum physics and a non-classical world. As it worked, the idea of a wave function—through common misunderstanding—was magically shifted from an abstract world of mathematical modelling, into the real world of physicality—falsely!

None of this should be confused with the idea of the dual wave-particle aspect of particles. This is the notion (several conflicting models move in and out of fashion, in fact) that particles can exist as both waves and particles. Some theories suggest there are no particles and no atoms at all: they are all waves!

When scientific exploration left the classical world and entered the sub-microscopic quantum one, it seems to me it also took a foot out of true scientific discipline and put it down in the discipline of philosophy, but fails to admit it explicitly.

Let me offer an example.

It concerns the experimental proof that, in some unexplained way, the human mind seems to influence the outcome of small pieces of matter and their events (the double slit detection issue).

Hugh Everett, a physicist in the 1950s, did not subscribe to the idea that consciousness had any magical powers in quantum mechanics. Because of the double-slit measurement problem and other similar paradoxes in quantum mechanics, philosophers are attracted by the idea of human consciousness collapsing a wave function. Better put—they started suggesting a human being is a deterministic presence in an uncertain universe. The consequence of such philosophic thinking is a centric one: human consciousness is the major actor in the universe, and without human consciousness, the universe would not exist.

A biggie, eh?

Everett was a materialist and a realist. He started work to calculate a solution to remove the weirdness of why it 'appeared' that mind affected matter. He applied an emerging new field of knowledge: information theory.

Everett's Ph.D. work provided an alternative quantum world interpretation (er... shall we call this philosophy instead?). Everett stated that for any composite system, say—a subject (person or measuring equipment) observing an object (a particle)—the statement that either the observer or the observed has a well-defined state is meaningless because the *observer* and the *observed system* have already entwined.

In such instances, we can only specify the state of one relative to the other; the state of the observer and the observed are correlated after the observation is made. This led Everett to derive from deterministic dynamics alone, the notion of a relativity of states, without needing to assume an abstract wave function or its collapse.

Everett was very knowledgeable about cutting-edge ideas when writing his thesis in 1954-1955. He took the view that information is physical, and he developed a mathematical argument showing how data correlates within itself.

This explains what happens in a superposition: a person looking at a gram of carbon existing in a superposition of a billion different places at once does not collapse any wave function.

To demonstrate the consequence of this mathematically, he came up with a solution showing that the observer correlates with every possible state of that gram of carbon on the pencil tip could be in. So before the human being looks at the gram of carbon, the carbon is in all the millions or billions or trillions of possible states, and after the human looks at the gram of carbon, *he or she* is in one state. In this theory, when the human-observer actually looks at the carbon pencil—or any other object—he or she splits like an amoeba.

In Everett's view, when the human correlates herself (interacts and exchanges energy with the gram of carbon), she splits into copies of herself, one for each element in the superposition—millions of copies. The observer is not split into replicates here in this universe. The observer is instantly replicated along with alternate replicated universes.

This split is what gives rise to the "many worlds" theory and the “many minds” theory.

The Many Worlds theory describes a series of branching universes constituting what has been popularised as the 'multiverse'. There are almost infinite copies of you, me, everyone else who was in the world at the precise moment you looked at the gram of carbon in those infinitely branching universes. There are branches where you have now died. There are branches in which you murdered people, became a dictator, a tramp, a coma victim, a sex worker, a pop-star. The reason for this is because each branched universe will continually split at every moment you exchange information with any other part of the system, and again in each subsequent system (universe).

Today, Everett's concept is one of several competing interpretations of quantum mechanics and the superposition problem. But right now, it's in lead place! The theory is well respected by high powered quantum physicists and mathematicians, despite there being some non-fatal errors relating to probability in his original papers on the work.

The critical, monumental aspect of all this which I should point out is that in this theory, everything which can possibly happen, will happen—every outcome of all decisions not taken as well as those taken. It suggests something else though, a truth through reductionism: you are actually living in your own universe, and I in mine!

If this theory is true, you are the central player in your own exchanges with the system you're in, and the ones that split from you. I am in yours too though, but I am not the central player. You are in mine but you are not the central player. We are all in a stream of our own universes playing out all of our possible actions and interactions with the ones we are in. If you ultimately die in one, another universe is immediately created for you—one where you never died. There are variations on the multiverse theory.

These ideas seem ludicrous to most people, yet all of them have substantial mathematical proofs supporting them. As I said, science has entered areas where it requires other more open-minded and less dogmatic professionals to pick up the baton, and look more into abstracts and philosophy to aid the painting of reality. When a few brave scientists do stretch out and push the envelope, they are often subjected to enormous pressure and ridicule—often by people far less able than them to grasp the subject coherently.

When people think they know something, often that which is taught and accepted without their own minds questioning what they are being told is true, they can be the most hateful lot in 'shouting down' any idea which challenges their knowledge. I wonder why?

Almost every teacher of Quantum Theory struggles with their explanations on the issue of wave function and its part in determining probability solutions to particle position within any given quantum system. Their students ask what exactly is Psi (symbol= Ψ), which is the maths icon for a wave-function of a particle or a system. Is it a true (real-life) attribute of a particle or a quantum system?

The problem the teacher has is explaining why the use of Psi, which is a brilliant and accurate 'device' for solving quantum systems measurement problems and predicting outcomes, actually works. The reason it does work has no logical, rational explanation. It is a made up notation derived from the mathematics used in solving normal wave problems in a classical world. Any frustrated student will soon learn what every other studier of quantum science learns, which comes down to the teacher's immortal response: "We don't know why it works, but it does, so don't worry about it. Just use it."

Which is all fine and dandy if your ultimate use of this scientific (mathematical) tool is in solving real world problems in your professional life and career. But, it is absolutely worthless to anyone wanting to try and picture or understand what reality is.

I have shown some diagrams below to help show the theories regarding multiverse theories (there is more than one). They might give a better idea of the extraordinary set of concepts involved. Of course, these are only theories, but their origin has been due to our attempts to understand the weird discoveries of quantum behaviour.

The multiverse below (1) stems perpetually from the original instance (A) —each splitting into two or more similar universes every time an action within the derived system is determined as having more than one outcome. A universe is created to resolve every possible consequence of that outcome, and so on for all derived universes.

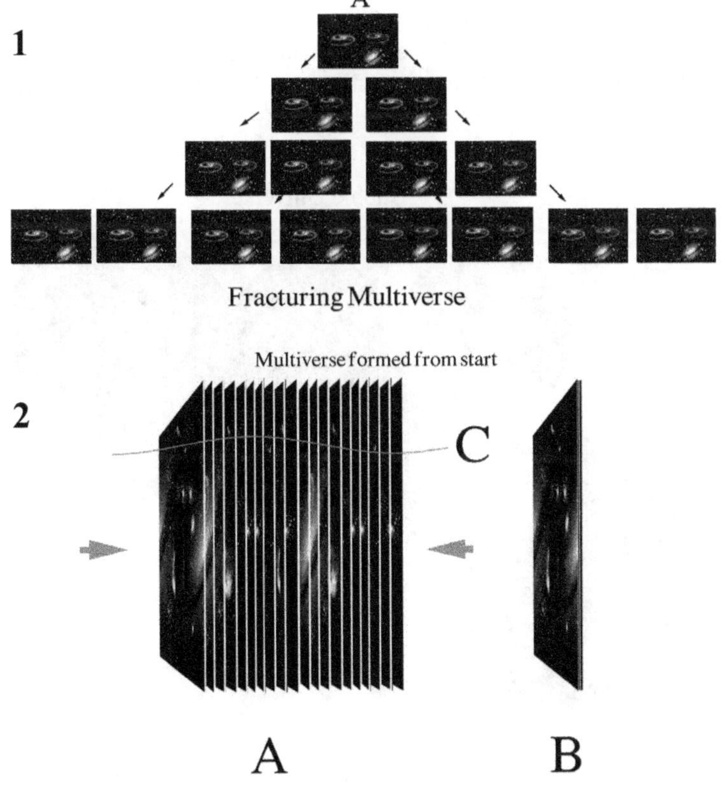

This leads to the creation of an infinite number of universes, all only very slightly different at the point of their creation. The multiverse (2), previous page, is an infinite set of universes (A) squashed together to occupy the same physical space (B). Interactions occur between sub-atomic systems and particles in each universe (C). This influences the behaviour of particles we observe, and our attempts to measure them at a quantum level, and thus provides a possible explanation of their weird behaviour.

Not so much a multiverse like the others which are conjectured theories to suggest weirdness observed in the quantum world, the bubble universe idea suggest many universes are formed or may still persist in being formed, each with different parameters and internal laws. (See 3 below). We can only be aware of our one and wonder why our universe can be so very finely tuned to enable the emergence of life. The odds are inexplicably against that. Such massive bad odds are greatly improved if zillions of universes are created but in most, the internal laws and subsequent internal systems are unable to support the emergence of life. This leaves us thinking we must be lucky if only one universe is created and the laws are so finely tuned to enable life. It's almost like it was made for us. But in truth, all these other universes exist, maybe almost infinitely, but are beyond our reality (outside our universe) leaving us unaware of them (according to The Bubble Universe Hypothesis).

Anyway, now we've had a glimpse at what the quantum world has to offer without the maths obscuring our attempts to imagine it, let's move on and have a look at reality and the impact of quantum weirdness in more detail. You might make a note to read this chapter again plus the previous one. I wasn't certain which order to put them in as they are... er... well if you will excuse the pun... entangled.

Bubble Universes.

3

* * * *

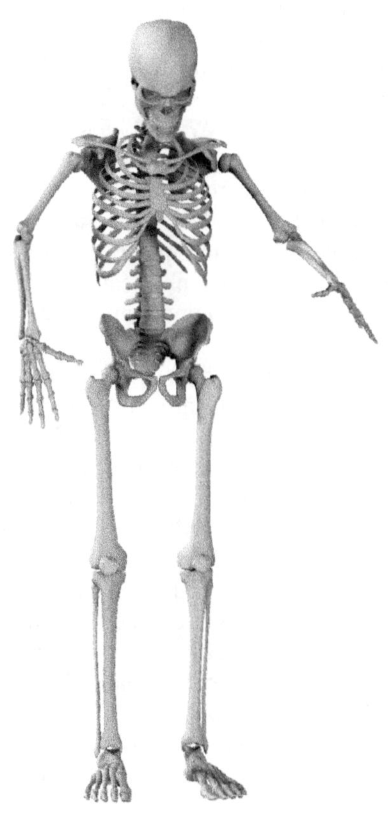

Who Am I?

Chapter 4: Who Am I?

Maybe I should really call this chapter, 'Where am I?' since it is really about the mind, our minds and their presence as sentient entities in this universe. We started out with a commonly understood premise: the idea that mind is a function or an attribute of a brain. It follows, as the brain is in our bodies—our heads—mind is exactly located in this universe in that part of you—your head. And my mind is in my biological head too.

The problem is this may not be true!

There is no apparatus discerned to be a mind anywhere. It's more a case of mind being a manifestation of a process, one which seems to use the organic structure of the brain as the place to present itself. This is not a semantic statement! We all possess a brain. We all believe through common sense, and by our physical senses connecting us to the external world, that we are within our heads and the physical world is, well... out there. Our senses provide information about our spatial position in an external reality. Our hands provide tools where we are able to feel and manipulate external objects. Simply put, our minds influence the external physical world through the actions and existence of our bodies, and our mind's presentation within the brain of that body.

In a macro classical world, there is no overwhelming body of evidence suggesting a mind can physically influence (say, move...) a physical object without using the body to move it. And just to be sure we are understanding things right here, yes—there does exist anecdotal evidence for 'special' people being able to move large objects (coins, table top items) by thought alone, but whether this is true or false—I can't do it, and neither can you. In fact, it appears almost no-one can.

This fact has a consequence. It is this. The physical world—the one we sense outside our heads—connects or interacts only with the physical body outside of our minds. Not outside the brain. If I accidently put a railing spike through my brain, it gets damaged, and a malfunction of some kind in my performance or behaviour or capacity to control my body will result! The physical world thus also interacts with the brain. The mind, however, only interacts with the brain. It—the brain—is an intermediary between mind and physical reality.

When someone dies, the brain dies. No brain means there is no intermediary for the mind to interact with and thus the mind de-coheres from the physical world: it loses its capacity to influence or interact with reality. Everyone recognises this to be a true and commonly observed fact. And because of it, and due to the gradual spread of materialism as a widely accepted philosophy of reality, it is also popularly thought that mind is the product of the brain. Which then leads to the notion that when you are dead, you are dead: no more 'mind' as it dies with you.

Now, the big million dollar question: is mind solely a product of the brain? We all know if you lob bits off the brain, the behaviour of a person alters. If a person's brain becomes drilled full of holes, as when people become sad victims of Alzheimer's disease, their minds cease to be displayed or presented to external observers (through the victim's behaviour) as 'normal', or rationally in contact with reality.

The logical conclusion then is *that mind is a product of the brain.* The network of electro-chemical activity across synapses between neurons is the manifestation of what we each internally consider we are—phantoms: electrical patterns mapped across physical structures. A very simple analogy might be we are the paint on a wooden plank, no wood—nothing to put the paint on. No wood: nothing for the paint to remain on.

All well and good... that is, as we've seen—if our brains are just quantum-less classical objects in a macro classical world!

What if they're not?

What if our brains also interact with reality at the micro level of the quantum world?

The human brain

Most of us are familiar with the brain's various functions. The one I am interested in here is computation or thinking. All the biological functions except for sensory perception to receive and store information are something else. Many people liken our brains to computers, but nothing is further from the truth. Yes, there are several analogies which may help us understand how we 'think' but there is a vital difference between a computer (most of those we currently use) and the processes going on in our brain. To think, it seems you need a physical network, where packets of information can be converted into chemical and electrical signals so they can sustain a complex web of interacting charges within a pattern. This pattern, this web of electrical charges, is, in essence, at any given time—you!

One might argue it is the locking together of chemicals related to those charges which is the biological 'thought' you. But it isn't. Consciousness is fundamentally an electrical state rather than a biological one, although in humans and other animals, an interdependence exists between the two; you're not dead when the chemistry fails: you're dead when the electrical activity ceases!

When we think more about certain things, the components of the brain (neurons, dendrites, and synaptic gaps) physically alter too. What you think, you become more fixed in. The biological brain shifts its physical pattern to enable quicker responses to repeated electrical stimuli. It creates new paths for information transfer, or shortens existing ones. It also brings into play more neurons to widen the 'processing' availability when you concentrate, or become deeply involved with a problem. Did you ever forget to breathe while focused on something?

If you want a detailed account of how neurons work, it's worth going to the wiki page here (*http://en.wikipedia.org/wiki/Neuron*) to find out. But here is a brief summary from the wiki site along with a diagram.

A neuron or neurone or nerve cell is an electrically excitable cell that processes and transmits information through electrical and chemical signals. A chemical signal occurs via a synapse, a specialized connection with other cells. Neurons connect to each other to form neural networks. Neurons are the core components of the nervous system, which includes the brain, spinal cord, and peripheral ganglia. A number of specialized types of neurons exist: sensory neurons respond to touch, sound, light and numerous other stimuli affecting cells of the sensory organs that then send signals to the spinal cord and brain. Interneurons connect neurons to other neurons within the same region of the brain or spinal cord.

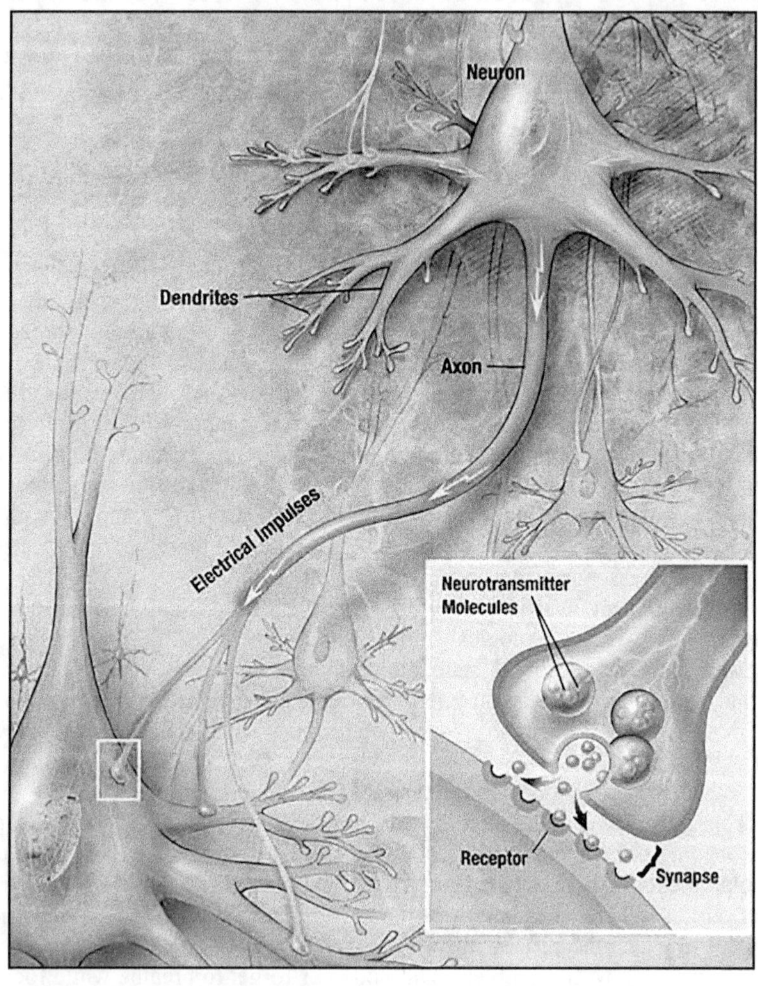

(wiki commons licence)
drawing illustrating the process of synaptic transmission in neurons
Source: www.nia.nih.gov Author US National Institutes of Health, National Institute on Aging

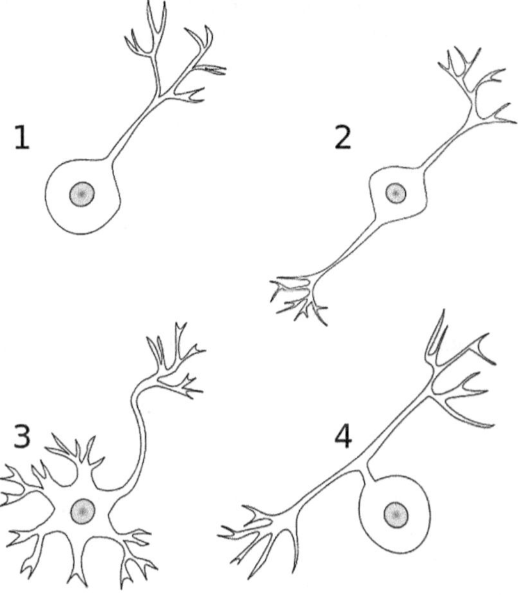

neurons drawing by Jonathan_Haas (wiki creative commons licence)
Different kinds of neurons:
1 Unipolar neuron
2 Bipolar neuron
3 Multipolar neuron4 Pseudounipolar neuro

A typical neuron possesses a cell body (often called the soma), dendrites, and an axon. Dendrites are thin structures that arise from the cell body, often extending for hundreds of micrometres and branching multiple times, giving rise to a complex "dendritic tree". An axon is a special cellular extension that arises from the cell body at a site called the axon hillock and travels for a distance, as far as 1 metre in humans or even more in other species. The cell body of a neuron frequently gives rise to multiple dendrites, but never to more than one axon, although the axon may branch hundreds of times before it terminates. At the majority of synapses, signals are sent from the axon of one neuron to a dendrite of another. There are, however, many exceptions to these rules: neurons that lack dendrites, neurons that have no axon, synapses that connect an axon to another axon or a dendrite to another dendrite, etc.

All neurons are electrically excitable, maintaining voltage gradients across their membranes by means of metabolically driven ion pumps, which combine with ion channels embedded in the membrane to generate intracellular-versus-extracellular concentration differences of ions such as sodium, potassium, chloride, and calcium. Changes in the cross-membrane voltage can alter the function of voltage-dependent ion channels. If the voltage changes by a large enough amount, an all-or-none electrochemical pulse called an action potential is generated, which travels rapidly along the cell's axon, and activates synaptic connections with other cells when it arrives.

The fact that it appears to be that we are minds, in brains, in bodies, walking and moving, interacting with a physical outside reality, is our first attempt to place who and where we are within the universe and within the system of reality. But, it is not the complete picture. Like our models and our diagrams of atoms in a classical-Newtonian picture of reality, it is only just a first step... a child-eye's view. It is merely a starting point enabling us to examine that notion in greater detail, and to refine our understanding of something so much more complex and profound. I might add at this point that neither science nor any other field of human understanding (philosophy, religion, imagination, art, mathematics, biology, genetics, music, etc.) can provide absolute answers or descriptions of mind, universe, or reality through its single isolated window.

In a similar way, no single sensory organ of your biological mass can define a complete picture of your external environment, nor a complete internal picture of yourself.

A novel way to consider 'self-awareness'

Many people think computers will one day become self-aware. The idea of intelligent machines—in this case, computers—becoming self-aware, or conscious through the complexity and number of processing units it has access to, is interesting. It's argued a similar process of increasing numbers of neurons and increasing neural complexity gave rise to a self-aware 'mind' in us. What I find fascinating is this. Both computers and our neuron brains exploit electrical effects to cross-communicate between processing parts and memory. It is easy to mystify the idea of 'mind' being the electrical network mapped across physical biochemistry (in humans) or silicon switches in computers. It is also as easy to make an error about the difference between computational thinking and aware thinking. The two are very different, no matter how many components, switches, or neurons are utilised in a system. The theory of humans developing self-awareness and intelligence is stated as being caused through the following: the increasing complexity of biological system required a control system to coordinate all the processes involved. As our bodies evolved, and grew even more sophisticated, so the number of neurons increased, and the interconnecting pathways in order to manage increased complexity. Ultimately, a threshold was reached where the increased numbers of neurons and number of interconnections triggered self-awareness.

I would suggest this is rubbish! It might be part causal, but its premise is based on entirely the wrong root, as I will now try and show.

Let's start with computers. Let us suppose the computer used neither light or electricity in its workings. For the sake of exploring the idea, consider a vast computer which is constructed purely of mechanical parts—-a super sophisticated Babbage engine. It would be very slow, I guess, clicking all those billions and trillions of micro-clogs around to process a program. Let's stretch the idea further and dream up the impossible idea that the mechanical parts can move so rapidly, their switching rates equal the speed of electrically powered computer semi-conductor switches: logic gates.

We are trying to remove the element of electrical or light energy being part of the computing process.

If we could now imagine increasing the number of mechanical switching units, possibly also making them ever smaller down into the nano-scale, would it give rise to a self-aware intelligence?

To me, the answer would be no. I might be wrong, of course. When considering the issue this way, does it suggest to you 'awareness' comes about through increased 'thinking parts'—the little cogs of the Babbage engine, or the tiny neurons of our brains? No, it doesn't.

So, we might all think intuitively instead that intelligence or 'self-awareness' comes about through the existence of an electrical network, both in the brain and in an electronic computer. We rightly or wrongly 'believe' electrical energy, when organised into a coded matrix of devices (either neurons, or solid-state transistors on silicon chips) is the magic ingredient holding the overall concept and 'feeling' of being self aware. But there's even more to this at another level, which offers a more profound introspection of intelligence and awareness.

Both neurons and silicon chips are state-changers. In computers, the introduction of electrical potential onto their semiconductor switches cause a change in the state of those switches. In neurons, a similar effect also happens (but not exactly the same). Both the biological devices and the non-organic devices have elements (chemicals) at an atomic level which are reacting to electrical charge and discharge. We might then argue, it is in the atomic nature of chemical effects and interactions where intelligence and self-awareness is being manifested, and not directly within the electrical matrix. If this is true, it suggests the super-duper Babbage engine can never exhibit true intelligence or be self aware simply because it lacks electrical, atomic, and chemical effects in the mechanical working of the device.

The implications of this type of thought experiment is we begin to suspect that 'self-awareness', as a product of high intelligence, is actually the result of many tiny bits of matter being co-ordinated into a unitary presentation of itself in a coherent way, but completely different from other tiny bits of matter around it (the unitary processing system). Yes, we are using electricity to affect the atoms and molecules; we are using matter to carry a charge; and we are facilitating communication between molecules through both chemical, and atomic influences caused by the electric force of matter. In thinking or processing information in both a computer and a human brain, we are 'playing' with and organising micro-bits of reality to accept a new role. They are no longer just carbon, silicon, or just atoms: they are now physical carriers of bits of intelligence working in unison to create a sum awareness.

To state it bluntly: intelligence, self-awareness, and thought is teased out of reality and the stuff it's made of. Consciousness is the unified presentation of atoms and electricity diverted from their original purpose and roles into a common one: that of holding and exchanging information about itself—the stuff of the universe. And, if, in truth, matter is our illusionary perception of a reality constructed of waves and frequencies, then consciousness is a special piece of that.

The illusion of a mind's spatial position

http://www.youtube.com/watch?v=mD7NzrBgXwM

It's possible to trick the senses in such a way that your perception of being in your head behind your eyes can be altered. Dr. Blanke, a Swedish scientist, has pioneered experiments using avatars, robotics, and 3D virtual reality techniques to demonstrate our minds can be 'tricked' into believing they are now in a physical space or body other than our own. Put simply, the subject is stroked whilst observing a dummy body away from them. The senses associate the stroking with the visual information such that within a short time, the subject associates strongly with being present in the stroked dummy. More sophisticated experiments demonstrate the only reason we feel we are in our own bodies is through processing tactile data (touch) with spatial data. What does this suggest?

Try this experiment for yourself...

Sit behind a friend. Use one hand to stroke your nose and the other to stroke your friend's nose, moving both your finger up and down synchronously. Make sure you keep your eyes closed.

Be patient. You need to do this for at least a minute and then a strange perception into your mind. You nose is either 3 feet long or it is your friend's nose!

The only reason you feel like you, your mind, is behind your eyes in your head has nothing to do with the mind being a component of brain in your head: it is entirely due to a combination of sensory data transmitted along your nerve cells from your eyes, and your sense of touch and balance. These senses can be used to transmit 'false' data very readily. Your brain is in your head, sure. But your sense of 'being'—an attribute of mind—and the location at which you site 'being' is entirely illusionary!

Let me demonstrate the consequences of this for a moment using a mind experiment. For this experiment, we need a head—one cut from its host body but kept intact by keeping it supplied with blood and oxygen. A science-fiction nightmare possibly. So, we have this head in a jar. Tubes run in and out bringing vital nutrients and oxygen via an external mechanical pump. We have removed the eyes from the head and also removed the inner equipment in each ear normally required for hearing. The nose and mouth have been completely sealed as the head has no need for taste, breathing, or smell.

We have discovered a way to directly link optic nerves to remote electronic cameras and auditory nerves to external microphones. But we are even smarter. The cameras and microphones for our severed head are placed in a helmet fitted to a person walking along a street in Australia. Our decapitated head remains in a jar, wired up to computers and all located in London. The internet provides the data carrier service between the helmet in Australia and the head in London. Where do you think the brain (mind?) in that severed head believes it is: in Australia or in London?

Real Mind

Taking this idea further, now imagine your true mind is one which for this thought experiment, I propose is not physically present in this universe, but is instead in a reality (dimension) outside of our capacity to realise. And let's say this mind, your true mind, is vastly superior in many ways to your 'known' mind. Of course, your superior mind must have some way to maintain a connection with your 'this-universe' one, and that connection is probably mostly a one-way only data traffic route—from known mind to superior mind. How could such a communication link be maintained? I propose (for the sake of our mind-experiment) the link is at the quantum level. The brain in your body has quantum world processors in microtubules which allow a form of data transmission from brain neurons to your 'other' mind in somewhere else.

Of course, this is an imagined scenario only. There is no scientific evidence supporting this imagined real-mind-elsewhere position or concept. It is a thought experiment. By the same token, there is no scientific way of disproving the notion either. It is an important idea none-the-less, not least because if such a situation were true, death of you is not necessarily death per se: it is simply a break in the connection, and an end of your real-somewhere-else-mind maintaining a sense of being here in this universe.

Maybe!

We can imagine one more step. Suppose the connection is not between one brain here and one mind there, but instead, many brains here and one mind there? In this instance, all connected brains here must die before the mind there can be cut off from all input here (in this universe).

But we are used to finding reasons for why anything should be a particular way. So, if we start to build an imaginary scenario, and we wish to apply rationale for why such a scenario could possibly have any potential of being true, we also need to look at why an elsewhere-mind needs to explore a different kind of presence in a different kind of place. One obvious starting point would be that the elsewhere mind has no way of experiencing what we here in our physical universe do as human being acting as self-contained entities. It cannot experience the complexity, mixed emotions, and detail of what we witness (feel?) competing with each other in a place which is imperfect—a place where pain and death are constant threats to our well-being.

One such place for evidence is in the multiverse theory we discussed in a previous chapter. In the many worlds theory, there are many 'yous'. It's unlikely that the human mind/brain duplicity is a classical behaviour organ only. Such a sophisticated and eloquent creation probably has quantum associated behaviours too. In the latest theory of the many worlds theory, it is proposed that interaction takes place between sub-atomic (quantum) systems in each universe. Unbeknown to our conscious minds, such interactions may be occurring all the time between all the different 'yous' too.

Such a scenario may give rise to special cases where such interaction is also an information one. Of course, this is a very huge conjecture on my part , but not impossible theoretically.

The boy who told who killed him.

http://www.amazon.co.uk/Children-Have-Lived-Before-Reincarnation/ dp/1844132986/

There are many anecdotal reports of people saying they can remember past lives. The following extract is from the book 'Children Who Have Lived Before: Reincarnation today' by Trutz Hardo

I have been a friend of the well-known Israeli doctor and Professor of medicine Eli Lasch for ten years now. He served for a long time as senior consultant responsible for the health services in the Gaza Strip and temporarily for the whole of the Israeli-occupied Sinai. Yet his own experiences gradually helped him to find his way back over the Kabbala to the inside, where he rediscovered amazing abilities that he had possessed in a past life.

After he had completed his career as a highly decorated medical doctor, he opened a practice in Israel where he worked as regression therapist and spiritual healer. In 1989 he came to Berlin where I took part in his seminars and where we ended up leading each other back into past lives. A few years later he appeared on television several times where he successfully conducted healings from a distance. All of a sudden he had become a well-known personality throughout the country.

In 1998 his highly interesting book appeared with the title From Doctor To Spiritual Healer. Eli told me several amazing stories about reincarnation, which had helped him to revolutionise his entire conventional doctor's way of thinking. In December 1998 I visited him in his flat in Berlin, where among other things he recounted the following event, which I will now tell you in my own words.

The Druse is a nation of approximately 200,000 people who settled in Lebanon, Syria, Jordan and the region that is now Israel a long time ago. They are neither Muslim nor Christian, for they have their own religion. In Israel they are mostly found on the Golan Heights. They are the only non-Jewish Israelis to serve in the Israeli army. Reincarnation forms the basis of their beliefs.

As soon as a child is born its body is searched for birthmarks, since they

are convinced that these stem from death wounds, which were received in a past life. If such marks are found on a child they try to discover something from his or her past life as soon as the child is able to speak in order to get the first clues to the circumstances of his or her former death. They are aware that small children often confuse past and present events and so experience everything as if it were the same life.

Therefore as soon as the child is three years old and is able to distinguish between events from the past and its present life, the child is taken to the place it has talked about and where it claims to have lived in a past life (provided that the child in question did mention such a place). Since this is usually a special occasion, a kind of native board of inquiry is formed, led by several respected village elders.

When a certain boy became three years old, on whose upper forehead a long red birthmark stretching to the centre of his head was found, a group of 15 men was formed. This group consisted of the father and other relatives of the boy, several elders of the village and representatives from the three neighbouring villages. From what the boy had said they were quite sure that he had lived in their immediate neighbourhood in his past life. Professor Eli Lasch was the only non-Druse who was invited to join this group because they knew that he was interested in reincarnation.

When they arrived at the first neighbouring village with the boy, he was asked whether it seemed familiar to him. He told them that he had lived in a different village, so they walked on to the next one. When they arrived there and questioned him again he gave them the same answer. Finally they reached the third village. Now the boy told them that this was where he had lived. All of a sudden he was able to recall some names from the past.

He had told them months ago that a man had killed him with an axe, but he had not been able to remember his own name and that of his murderer. He now remembered both his first and second name as well as those of his murderer. One of the oldest people of this village who had joined this group had known the man whom the boy named. He said that he had disappeared without a trace four years ago and had been declared missing. They thought he must have come to some harm in this war-torn area, for it often happened that people who strayed between the lines of the Israelis and the Syrians were taken prisoner or shot if suspected of being spies.

They went through the village and the boy showed them his house. Many inquisitive people had gathered around. Suddenly the boy walked up to a man and said, "Aren't you ... (Eli forgot the name)?" The man answered yes. Then the boy said, "I used to be your neighbour. We had a fight and you killed me with an axe." Eli told me how the man had suddenly gone white as a sheet. The three-year-old boy then said, "I even know where he buried my body."

How could he have known where his former neighbour had buried his body after his death? Almost daily, my clients describe to me the following post-mortem scenario during regression therapy: after death the soul leaves the earthly body and in most cases is able to see the body from above. Often it hovers there for a while and can see exactly what happens to the body. We will hear more about this from other children later on in this book.

Some time later the whole group followed by many inquisitive people were seen wandering off into the nearby fields. The man whom the little boy had recognised as his murderer was asked to come along. The boy then led them to a particular field and stopped in front of a pile of stones and said, "He buried my body under these stones and the axe over there."

They now removed the stones and underneath discovered the skeleton of a grown man wearing the clothes of a farmer. A split in the front of the skull was clearly visible. Now everyone stared at the murderer who finally admitted to this crime in front of everyone. Then they went over to the place where the boy said the axe was buried. They did not have to dig for long before they held it in their hands.

Reincarnation is a fact of life for the Druse; they need no proof to secure this belief; and yet it always amazes them every time reincarnation reasserts itself in cases like this one. The Druse also believe that they are always reborn as Druse. Perhaps group regression among their people would prove whether this statement is true. Eli then asked the people what would become of the murderer. They said they would not hand him over to the police, but they themselves would decide on an appropriate punishment for him.

True or False?

Of course, we are left in a position of not fully accepting this account to be true because we have no immediate way of checking the evidence. There is also no way of re-establishing an experiment to provide any evidence of reincarnation. What we do have though is a theory of a multiverse and a multi-you. Maybe there is also a 'master-you'—a kind of consciousness collecting all your multi-experiences in all the universes. Maybe information from the murdered you from one universe is somehow passed to the just-born you in another universe? This doesn't actually solve the mystery of the boy who believes he is resurrected.

The enigma is only solved if there are also multiple existences of you being born and dying at different times throughout all the universes. And why not? If a reality is capable of coming into being such that there are multiple copies of itself all proceeding in various and differing evolutions, then why should it not also speed up the novelty-capturing process by ensuring different versions of you are born at different times or even at the same time in each of the different universes?

Too far, too soon? Right! Maybe I have lost my credibility by going too far with philosophical conjecture too quickly and we find ourselves moving away from the pool of evidence science has amassed. Let's take a different look at the problem using a model closer to home. The humble computer!

A thought experiment

For the sake of humouring me, and to possibly understand how difficult it actually is to understand exactly who you are, imagine for one moment you are not a human being. Instead, you are a sophisticated piece of coding running at rapid speed along the bus wires of a computer. Thirty-two bit computers have 32 such lines for the greater part of the equipment, allowing 32 simultaneous bits of code to move along the wires/ribbons at any given time through ports, into registers etc. You might be made up of several million bytes of code. This is you. Meanwhile, you co-exist with many other byte–program people, all competing for the bus bars, the registers, the processors and the ports. Some of you byte-program people are smart and curious. Some are just happy being bytes running along the wires and never worry about where they are. But, you, and several of the other 'curious ' byte-people begin to collaborate.

Over a period of time, you learn much about the coding world you live in. You discover discrete timing pulses drive your very existence. You realise you appear to be made of organised bits of energy (electricity), you discover each of you have similar stings of code within your different coded selves (genetic coding?), and you realise that something is not quite what it originally seemed to be. For one thing, you discover your reality has areas which seem to concentrate all activity like it has an overwhelming gravity type pull (ports). All code-people seem to only last so long and then disappear (die). Now and then you are collectively able to discover some chunks of code you encounter are not byte-people at all: they are communication code strings being used by the system to keep everything running and synchronised, but you don't know about the system.

It is impossible for you to 'see' the bus bars, the wiring, the semi-conductors and electronic digital chips you inhabit. You will never be able to prove those 'solid' things are there. How can you? You are code. All you see is code. You can conjecture that you may well be code inside a system you can't fully realise, and you can 'dream up' all kinds of things which that system might be. But never, ever, will you be able to find any evidence of the physical structure of copper and silicon.

This is the same issue you have now. You are a mind inside a biological system inside a reality system, and as you are inside—not outside—and made of inside stuff, you will never be able to prove where you are or who you are... not really!

What you are left with, if you do wish to try to understand the possibilities of who and where you are, is but one thing: imagination. Now, this is not wild, dream-up-anything kind of imagination, but the type which somehow seems to fit together the weirdness we now understand about our reality as we perceive it to be.

To understand reality and who you are (what you are?) within it, you need to think like my imagined byte people. You look at the evidence you can determine, and even though you can't get any evidence about things lying outside our local reality, like the bus bars in the computer, you conjure up a best fit with your imagination. Science does that! Philosophy does that too!

Computer simulation

I am not suggesting we—you, I, us—are part of some grand computer simulation, although that is actually possibly. What I am suggesting is we observe many things which are dismissed too readily, or are explained too quickly as aspects of people suffering delusional thinking or mental disorders.

And, of course, in an anecdotal world and wherever it is difficult or impossible to set up faultless, repeatable experiment and demonstration, it is extremely difficult to sort the mad from the sane. But it doesn't mean everything reported and untested is not real.

People

As a human being on a planet of nearly seven billion similar people, you actually share very little in common with any of them. From birth until now, you have lived in different places, and witnessed completely different events. It is likely though you will have witnessed a variety of similar emotional events and feelings: sadness, joy, grief, love etc., But this does not make you like them at all. Remember, you are a mind and that mind, your mind, has built its own internal model of the reality outside. And so has everyone else. Seven billion people: seven billion mini-universe models in 7 billion brains and each one uniquely different to every other one. A multiverse in flesh, no less!

The question of who you are is a compound one. You are the product of all the experiences you have had so far and how many of these affected how you were originally conditioned in your growing years. But you are also more than this. You are part product of all the other people who have exchanged information with you, through physical interaction, through books, films, TV, story-telling, music etc. Not just people alive today, but people who are long dead but who left a legacy of themselves in the things they created which survived their physical deaths. You have encountered past peoples experiences this way on your journey.

You might be more than this still. You might be carrying within you access to the sum total of all human experience and unknowingly tap into that resource too without realising you are doing that. Most of the time, you are concerned with the 'now'—a sensation provided by just a thin layer, a spray paint thick skin of brain over the frontal lobes. This, I call your 'now' engine. But 'its' communication with the deep sea of memories and processing capability of your huge greater mind—the billions of neurons, dendrites, the multi-billion synaptic gaps, and ever holistic mysterious brain in which your entire universe has been constructed—is limited and almost only one way. You build a universe inside, but you never really see the whole of it.

Something to ponder

Consider the two images on the right. The first is a computer simulation created from information we have regarding the distribution in space of visible matter within the dark matter. The second image is a computer simulation of the neural network in a human brain. Similar pattern, eh? It makes you think doesn't it? Could our own universe be a thinking neural network?

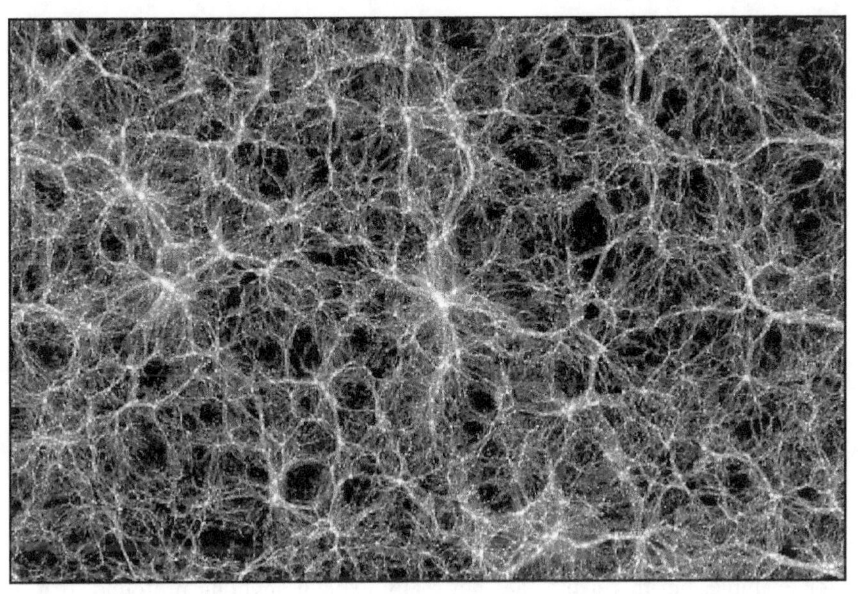

Compare the image (distribution of visible matter in space) above to the one below (a neural network).

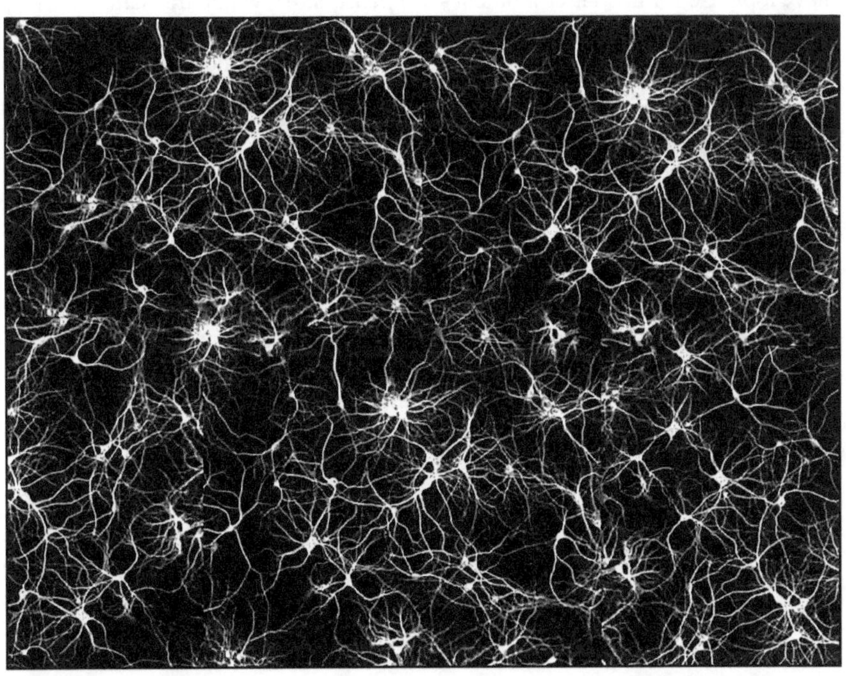

Chapter 5: Clues in our biology point to our profound identities

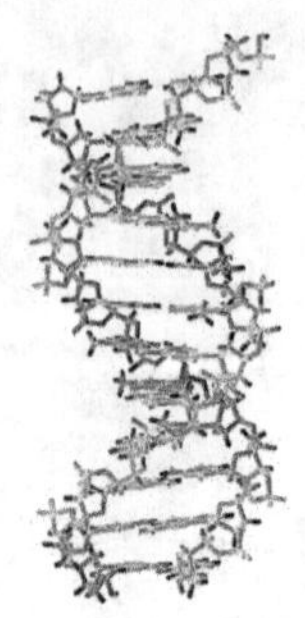

If you are going to make a machine to perform a job, you need to write out a list of functions you wish it to perform, and then design the mechanisms to carry out those functions. Like-wise, if you have a machine and wish to consider what thinking went into its design, you have to reverse engineer the machine. Maybe within our human experience as biological forms, there are clues which indicate we do not only exist as minds here in these brains, but as minds of a different ilk interacting with the presented mind of a biological brain.

One of the most unexplained aspects of our biological machines is the one concerning pleasure. And of this—sexual pleasure. Scientists can nominate a family of chemicals along with areas of the brain they affect during orgasm, for example, but this itself does not really explain anything profound. It just explains what the functioning mechanisms are. As we appear to be designed and created through, and by, the universe itself as part of a conjectured evolutionary process, it should come as no surprise this all leaves me wondering where the idea of bliss originated from. There is a moment during sexual orgasm where an intense high is experienced followed by a wonderful blissful, almost thought-free, feeling of peace and warmth. The French even have a phrase for this latter moment—***La petite mort*** which translates to "The little death".

Let me first try and explain this feeling. Imagine no pain, no fear, no concern to anticipate any immediate or long term future. Imagine being filled with a warm calm, almost as if you were no longer in a body built to interact with a physical reality, as if, instead, you were just in a state of being just that... being. No analytical mind, no problems to solve, no nerves to signal pain and discomfort, no future, no past, just this one warm 'now' moment. I'm sure you know the feeling too. And we all desire to experience this over and over again throughout our lives.

It's easy to see such a reward might be on the cards for performing an act which in the past was very likely to bring about the creation of a new life. We human beings though, are a clever lot and we've uncovered ways to experience sexual-bliss reward without paying the full price of having a child and looking after it for eighteen odd years. Nonetheless, we could ascribe to the universe an attribute of being a clever designer to dream up such a unique reward encouraging us to seek it through sexual reproduction. What we must remember though, is that all too readily we attribute almost God like cause and effect to a universe which science clearly has no proof or suggestion of it being sentient. But we do it all the time.

What I find to be an enigma, and possibly a clue to a bigger reality being in effect than the one we are capable of perceiving, is where has this experience of bliss been defined from? To explain further—how can a mechanism evolve a system within our biology where such a system seems to create a unique but temporary state of mind which itself is almost no mind? In post orgasmic bliss, our normal every-minute thinking ceases. Instead, we experience a non-thinking reality almost completely internalised and with no real interaction with the external physical world:

"Right", says the Universe, "I need to ensure these bipeds reproduce. Um. They're not stupid and they know they'll spend most of their time running around looking for food and water to keep themselves alive. They are just not going to want to reproduce knowing they'll have to run around a damn sight more to keep other hungry mouths fed. What can I offer them they can't refuse? "

Think. Think...

"I know. I'll offer them a brief period of being without care or needing to anticipate the immediate future. Now, where can I draw inspiration from to create that feeling."

Think. Think...

"Of course. Silly me. I could give them a few minutes of this. I'll give them a little death!"

Do you see my point? Here we have a chicken and egg situation—how would an evolving biological mass know what a blissful experience is?

There is a solution to the enigma. Bliss is a pre-conceived idea! In this solution the notion of bliss is already a known experience simply because it is the natural state of a sentient existence which does not need to anticipate an unfolding and constantly changing physical reality. It is the natural state of a mind which does not have to interact with matter and energy as we perceive it in our material universe. But is there a clue here about life and death itself? I believe there is. If each of our minds can experience a state of bliss, even for a few minutes, it means such a state of existence is possible. Of course, our bliss seems to take place inside our brains as a product of chemicals like dopamine being released in abundance and changing the state of the neuron firing pattern.

I will not refute it to be a brain organ experience (for the sake of argument) here, although I have deeper issues about that. What I will say though, is during that brief blissful state, not much else of what we call the conscious part of the brain appears to be doing anything. Neurons are not firing up to communicate with the unconscious part of the brain to store and retrieve memories, or indeed to calculate if that noise you might have just heard downstairs is in fact an axe murderer breaking into your home. The brain, and thus the part of your mind seemingly residing in it, no longer cares about the real world at all, nor does it care about the universe it's built within. It, you... well... you just want to stay in that 'preferred' state, don't you?

If say, at that precise moment of full bliss, you became permanently cut off from all others areas of the brain such as memory, hearing, sight, taste, smell, touch, the ability to count, spell, read, recognise patterns—it would make no difference to you what-so-ever—not until the bliss ended, of course.

So, here we have a clue suggesting you don't need all the extra weight of biological grey mass to experience being at all. You don't need eyes, teeth, tongue, lips, skin, ears, nor a body—other than those vital parts keeping the oxygen-rich blood flowing to the areas of your brain now presenting the experience of dopamine-induced bliss. If we could just map that state and just those neurons firing, or other components representing their connecting pattern and electrical presentation, onto either a biological or non-biological system, would you still be? Sure, you may not be the whole you, the one with history and a sense of being involved with a journey in physical reality, but it seems to me, you will have the experience of being. In fact, providing the state of bliss is maintained, your experience of being will be a whole lot nicer than your present physical one: no wants, no desires, no motivation, just being and feeling good!

Does such a scenario already exist, but instead of you being mapped onto other components of this universe/reality, you are already mapped onto the other states of matter in another type of reality—one which does not present itself as a physical universe emerging from a singularity?

The fact your mind can be pushed into a desired state of being—one unlike any other you are likely to love as much—makes me wonder if that is, after-all, our natural state of being, and this one where we run around trying to achieve goals, is the 'unnatural' one. If we had no physical bodies to look after and feed, we would have no need for most of the activity of the brain, nor the part of mind manifested to carry out and worry about such activity.

Death itself brings about death of the brain. That's a whole lot less neurons firing, and a whole lot less brain activity than would be going on in your normal living state. But it is also a lot closer to the amount of neurons firing, and brain activity, in your blissful state. It would be tantalising to say dying seems to take you closer to the physical bliss status: less neurons firing, less concern about interacting with the external world.

Many people *do* report experiences of dying. Of course, they may have only died briefly before being resuscitated, otherwise they would not have any physical presence here to report back on what it's like to die. Such experiences are termed NDEs.

Reports of death experiences (maybe?)

(NDE) 'Near Death Experiences' is a phrase first used by the psychologist Raymond Moody in 1975 to describe patient accounts of what he believed was 'a glimpse of the beyond.' Common experiences are reported by anywhere between 12 and 18 per cent of survivors of cardiac arrest, of seeing tunnels, bright lights, mystical encounters, dissociation and detachment from the body.

Scientific studies have conjectured these experiences to be the result of various elements in the neural network (brain), being flooded with endorphins. Science considers the NDE to be a product of chemical transformation of the brain and not a clue to the existence of mind existing beyond the life of the brain.

However, an existing assistant professor of medicine at State University, Dr. Parnia, reported to have found no biochemical evidence supporting the role of commonly speculated biochemical changes in near death experiences. In a BBC documentary titled The Day I Died, Parnia expressed his views that the mind was possibly a phenomenon that could exist independently of the brain.

In a new study just recently published in his book 'Resuscitation', Parnia released findings from a four year study involving 2060 people in the UK, Austria and the US who experienced a cardiac arrest and were successfully revived. The statistics of the study reports that 40 percent of revived patients recalled a form of awareness they attributed to the period they were deemed to be clinically dead. Their narratives not only described the bright light, and mystical encounters, but also describe the details of how various machines sounded, and the actions and conversation of particular medical staff.

In itself, this data is impressive in its scale and detail. Parnia methodically dismisses the likelihood of the usual biochemical explanations. For him, the brain is just too dead to do much of anything while the heart isn't pumping. The results suggest awareness during cardiac arrest can't be compared to that during anaesthesia, for instance. Thus, within a model that assumes a relationship between cortical activity and consciousness, the occurrence of mental processes and the ability to accurately describe events during the period prior to resuscitation—-a period when cerebral function is absent or at best severely impaired—-is perplexing. In a 2013 interview, Parnia responded to the fact his claims verged on the supernatural by saying "Just because something is inexplicable within our current scientific understanding, it doesn't make it superstitious or wrong. When people discovered electromagnetism, forces that couldn't then be seen or measured, a lot of scientists made fun of it. It could be that, like electromagnetism, the human psyche and consciousness are a very subtle type of force that interacts with the brain, but are not necessarily produced by the brain."

Such findings are difficult to re-establish experimentally. Most people experiencing brain death due to cardiac arrest, are not really in a position to be monitored for discrete brain activity when just about the whole medical staff are rapidly engaged in trying to kick-start the patient's heart, and prevent permanent death. Also, few people are likely to queue up as volunteers to have their heart stopped for a minute or two before the medical team attempt to get it going again, often an unsuccessful task!

But we now seem to have two clues (not proofs) that mind may not altogether be a complete manifestation of a brain. Yes, a brain is certainly required to negotiate a physical reality, but is it necessary to experience a state of being itself, one not founded in the physical reality we come to know of in our physical lives?

One thing which always amuses me is the way many dogmatic scientists (not all are), present every human experience as a manifestation of malfunctioning brain mechanisms. Schizophrenia, seeing ghosts, having visions of the future, meeting angels, being abducted during sleep by aliens, possession, talking to the dead, glimpsing heaven... all are quickly dismissed and blamed on areas of the brain working differently, or chemicals there-in being imbalanced. The thing amusing me is this: how do you know your brain is working properly at all? For all we know, we might be the surviving remnants of people whose brains were wired differently to ours, and they were far more in contact with all of themselves (mind in brain here, and a possible distant mind not in physical reality); maybe we, their descendents are limited in this capacity. We have lost the connection with our other mind. We just think it's only this here, our universe, our physical reality, our brain, our brain death, our mind death. We just use the words 'normal working' to describe an almost entirely subjective perception of just what normal is. No-one is doubting most of our experiences as human beings is largely due to our brain activity, but there is no absolute proof available for mind being only the product of electrical patterns existing within the brain. And yes, I do realise that because science can't prove something to be absolute, it doesn't mean the alternatives to the scientific *beliefs* are any more right or true.

Some things are always likely to remain outside scientific provability due to ethical or moral considerations. You can't really go around killing people and monitoring every aspect of the dying process before attempting to bring them back and report on their experience.

Time to die

Time itself is an elusive thing. You can fall asleep for a few minutes and yet awake and recount dreams which seem to last hours or days. Wait for a bus to attend an urgent appointment, and ten minutes seems like an hour. Go to the cinema and watch a movie you love, and three hours seems fleeting.

So what exactly is time? Is it real? The following extract from Wikipedia summarises an answer:

Two distinct viewpoints on time divide many prominent philosophers. One view is that time is part of the fundamental structure of the universe, a dimension in which events occur in sequence. Sir Isaac Newton subscribed to this realist view, and hence it is sometimes referred to as Newtonian time. An opposing view is that time does not refer to any kind of actually existing dimension that events and objects "move through", nor to any entity that "flows", but that it is instead an intellectual concept (together with space and number) enabling humans to sequence and compare events. This second view, in the tradition of Gottfried Leibniz and Immanuel Kant, holds that space and time "do not exist in and of themselves, but ... are the product of the way we represent things", because we can know objects only as they appear to us.

So, we can't all agree whether or not time is a real trait of our physical universe or not. I think what we can agree on is that our sense of time is entirely subjective. Some things seem to pass quicker than we thought, and others seem slower.

How long does it take to die then?

Or more importantly, when, at what point, are we truly dead?

This is a tricky question. Is it when your heart stops working, (a physician's answer), when your soul leaves your body (a religious person's answer), or when a certain part of your brain stops working (a brain surgeon's answer)? Who decides when you're dead? Can you be dead in body, but not in mind? A study published in a 2010 issue of the New England Journal of Medicine adds intriguing thoughts to this often hotly debated topic.

Typically, a severely head-injured patient is checked for consciousness soon after his or her accident. The doctor will look for patient's ability to track a moving item with the eyes or respond to a question by moving a hand or a finger when asked to do so. The doctor 'looks' for a sign that the patient is aware in some way or another (often in a basic state) to respond intelligently to an external stimulus or question. If there's no response and 'apparent' unconsciousness continues for any persistent length of time, the patient is considered to be in a persistent 'vegetative state'. Note here... 'vegetative state'. However, it doesn't mean such a state will go on persisting. Sometimes, inexplicably, the patient will recover and re-assume a conscious state. The problem here is not so much that the brain has ceased working in its entirety: it is still maintaining all of its essential control systems of the body... heart, lungs, etc., It just appears that if a mind exists still inside that brain, it does not present itself.

Monti and his colleagues in Belgium have added a new test for consciousness, applied to over fifty people in a proclaimed vegetative state. Using an MRI machine (which monitors for active neurons in the brain), Martin Monti and his team monitor these patients' brains when they are asked a question. And, amazingly a handful of the patients' brains light up 'Yes' or 'No' just like your brain or mine would if we were asked a question. These people are thinking—they are responding to a specific question. They are not vegetables after all!

One might ask, what are these people experiencing, if anything, while in such a state? It appears to me the whole question of consciousness is a tricky one indeed. If consciousness ceases, such as partially happens when you sleep, the mind still goes on working in a working brain, you are just not aware of it, not unless, that is—you wake up again, at which point, you may report dream-like experiences. When you die, in most cases, it appears it is not a sudden event but a progressive one where people slip into ever deeper type comas as their brains become starved of the oxygen it needs for survival. At some ill-defined point, every single neuron in the brain ceases to function. And at some later point, every component once deemed a neuron is no longer anything but a decaying chunk of grey slime.

From a purely scientific materialistic stand-point, one could say you are dead when insufficient neurons are left firing in a large enough pattern to present you, the mind within, with enough 'narrative' (be it pictures, sensations, or a dream-life awareness) to have a coherent experience. It could be an entirely imagined experience, but as long as it is something, you, the mind, you are still here and part of this physical reality—even if you are no longer able to interact with it externally.

Now, this is where the subjective question of time comes in. If the last few thoughts you ever have with dwindling activity in the brain is a dream of going through into a bright light and meeting God, and then too few neurons persist firing to maintain the pattern, the amount of time you are engaged with the dream may 'feel' like forever. And this may happen. You may dream up your own heaven and go there simply because it is the last pattern of brain activity before insufficient activity stops the dream from progressing. To you, it is a forever moment. To us, you just died!

This is important. Since you are a mind which appears to need a physical structure to map itself onto to be here in our reality, when the physical structure is no more—you still persist because your last thoughts were timeless. You exist in an unending dream because time itself was an illusion in the first place. No time: no moment when the dream ended and darkness came. Mind carries on with no physical structure at all. It carries on within an abstract notion outside of physical reality. Immortality and un-changing sense of bliss is achieved!

However, that is an entirely scientific materialistic answer which may satisfy all who believe mind is just the result of a brain. I don't actually believe that is true.

More clues: species memory

One of my favourite champions of proper inquisitive and open-minded thinking is Rupert Sheldrake (*http://www.sheldrake.org/*). Sheldrake was born in Newark-on-Trent, Nottinghamshire. His father graduated from Nottingham University with a degree in pharmacy and was also an amateur naturalist and microscopist. Sheldrake credits his father with encouraging him to follow his interest in animals, plants and gardens.

Sheldrake says, "I went through the standard scientific atheist phase when I was about 14. I bought into that package deal of science equals atheism. I was the only boy at my high Anglican boarding school who refused to get confirmed. When I was a teenager, I was a bit like Dawkins is today, you know: 'If Adam and Eve were created by God, why do they have navels?' That kind of thing".

At Clare College, Cambridge, Sheldrake studied biology and biochemistry, and after a year at Harvard studying philosophy and history of science, he returned to Cambridge where he gained a PhD in biochemistry for his work in plant development and plant hormones. He worked as a biochemist and cell biologist at Cambridge University from 1967 to 1973 and as principal plant physiologist at the International Crops Research Institute for the Semi-Arid Tropics until 1978.

Author, public speaker, and researcher in the field of parapsychology, he is known for his "morphic resonance" concept. Conceived during Sheldrake's time at Cambridge, morphic resonance posits that "memory is inherent in nature" and "natural systems, such as termite colonies, or pigeons, or orchid plants, or insulin molecules; they inherit a collective memory from all previous things of their kind". Sheldrake proposes that it is also responsible for "telepathy-type interconnections between organisms". His advocacy of the idea encompasses paranormal subjects such as precognition, telepathy and the psychic staring effect as well as unconventional explanations of standard subjects in biology such as development, inheritance, and memory.

Morphic resonance is not accepted by the scientific community as a real phenomenon.

So let's take a look at this idea of species memory. First of all, what exactly is morphic resonance? It would be wrong of me to lift entire tracts from Sheldrake's many books, which I would recommend to every reader (references in appendices). However, I can summarise some of his ideas. Most people, for example, consider they have inherited their characteristics and traits through their genetic coding, a legacy of the mixing of their father and mothers genetic code. However, most of this belief is unfounded. So far, very few existences have been discovered for gene sequences which, say, predict susceptibility to certain diseases or indicate a predisposition to certain biological weaknesses in any human being. The same is true for many seemingly inherited characteristics in all living things. Whereas genes held out the promise to be the code of life, a code which was thought once understood, would predict from an embryo's genetic mix, the future disposition, traits, and strengths of the child to be—no such consequence has happened. Yes, many traits are inherited from the genetic coding, but it seems we inherit other traits from our ancestors, yet they have not been discovered as being present in the genetic coding.

So where does this information or influence come from? Rupert Sheldrake proposes that every species has a species collective memory. He does not suggest this is something like a load of data on a hard disk in the sky or in some paranormal space. He suggests a kind of resonance exists (me: at a quantum level?) where each new member of a species is informed and aided by the species memory. Where and how this resonance exists is not determined.

It is this resonance which informs a horse embryo it is a horse being built. It is this resonance that informs fingers being built on a baby's hand in a human womb, they have reached their right length and number of digits. It is this resonance that informs birds in England that birds in Germany have learnt to peck through aluminium milk bottle tops left on doorsteps to drink the cream on top of the milk, so the English birds immediately start doing it.

Sheldrake not only has ideas about morphogenetic fields but ways of testing his theories. The theory is: morphogenetic fields carry information only (no energy) and the fields exist throughout time and space without any loss of intensity after they are created. They are initiated by the patterns of physical forms (both organic and inorganic structures: cells, molecules, crystals, animals, birds etc.). Once created, the fields help to inform later similar systems. A newly forming system (baby in the womb, a new crystalline structure) "tunes into" a previous system by having within it a "seed" that resonates with a similar seed in the earlier form.

In this theory, DNA in the genes of a living system does not carry all the information needed to shape that system, but acts more like a "tuning seed" that tunes into the morphogenetic fields of previous systems of the same type.

There are many mysteries concerning current scientific knowledge of human memory. Many people have been discovered with extraordinary brain defects where most of the core of the brain is missing and yet the person does not exhibit any adverse effects concerning IQ and memory. In effect, our brains seem to be very difficult organs in which to pin down where memories are actually stored. Large sections of a brain can be cut away removing memories, but somehow, inexplicably, those memories return. It is as if memories are 'stored' holographically across the entire neural network, or—as Sheldrake's theories suggest—they are not stored in the brain at all, but in a species collective memory which the brain tunes into.

If these ideas are correct, then the "storehouse of species memory" is not private since morphogenetic fields are universally available and continue to exist regardless of what happens to their original source. The thing that makes our mental processes seem private and our personal memory rather than someone else's is that we naturally resonate most strongly with our own past mental states. In other words, each of us broadcasts on a unique channel to which, generally, no one else listens. Someone else could tune into your memory and thoughts, and indeed, in practice, we do—as the common experience of "reading" another person's mind attests to, or possibly people who say they remember past lives (*the morphogenetic field reaches back and forth in time itself*).

Well, that's the theory anyway. What proof is there that these ideas have any validity? One of the most convincing proofs involved teaching rats to run a particular maze. Each new generation of rats learned it faster even though there was no direct physical way for any generation to pass its learning on to the next. Since then, a variety of new experiments have been performed. Although none of them has provided an absolute proof, they throw up results which are difficult to explain away using normal scientific cause and effect. One experiment carried out in Britain at the Open University involves the humble fruit fly, although the experiments were not carried out to test any of the morphogenetic fields theories.

The experiment was studying the effects of ether on fruit flies. If the eggs were exposed to ether three hours after they're laid, some of the flies developed abnormally, with four wings instead of two. Normally, the fly has two wings (a pair) and a pair of balances called halteres.

As they went on exposing subsequent generations of flies to ether, the abnormal proportion increased more and more. Then they took flies from the basic stock which had never been exposed to ether—neither they nor their ancestors—and exposed them to ether. They got a much bigger response of deformed flies than at the start of the whole series of experiments. In other words, the treatment of the other flies with ether seems to lead to a bigger response in subsequent (but non-descendant) flies treated in the same way.

Rupert Sheldrake's theories are hotly refuted by many conventional scientists, even where they are unable to offer proofs of alternative solutions to some of the mysteries he discovers, and illuminates as existing within conventional, ever-more materialistic and ever-more dogmatic scientific disciplines. But an increasing number of scientists are taking his ideas seriously. If further experiments continue to produce significant results, it will soon be the critics who will need to prove their case.

What appears to be the difficult thing for materialistic scientists to accept, is the idea that information, created by and retrievable by physical systems, can be transmitted and stored in a non-physical form. To them, information must still be carried and retained on bits of matter: photons, neurons, silicon chips). It contradicts the materialist vision that has served as the basis for most of the sciences, and many (but by no means all) scientists are reluctant to let go of that vision. Yet the experimental support for materialism has been crumbling since quantum physics arrived on the world stage, bringing with it "instantaneous communications", implied by Bell's Theorem in quantum physics, and 'entangled particle systems' along with light behaving as matter, in the guise of particles, and as energy waves.

Morphogenetic fields may also help to provide mechanisms for all kinds of human experiences which are scorned at by conventional science without any real comprehensive research ever being undertaken: telepathy, memory of past lives, meeting dead loved ones etc. If indeed, the human race has a collective memory, where all of our experiences are stored, some people's ability to tune into that massive resource, may be more enhanced than others.

It seems to me, millions of people throughout the world experience things not readily explained by current science narratives and theories. Yes, many of them might say things as the result of being delusional or having brain damage, mental illness etc, but like all systems of data collection—buried within the data there will always be exceptions to what can readily be explained. In the countless reports for hundreds of years of so called 'psychic phenomena', very little of it has been properly investigated because science cast it into the rubbish bin without even removing the wrappers.

Death to religion, long live science

Science struggled against religious beliefs in the nineteenth century. But in the 20th century, overtook it as the western world's number one belief system. And just like people in the past were once convinced of Odin's, Zeus', and Vishnu's existence without question, today they listen to the priests of science who are equally guilty as priests of all previous faiths of 'bending' the results to fit their own ends.

Science is no longer the open-minded enquirer, championed by unbiased curious minds looking at our world and our experiences, it has become a brotherhood of often spitting adversaries, camps of close-minded people, defending their own research, scorning other camps and their findings, and coming together as one united brotherhood, only to throw ridicule on any suggestion that other explanations exist for the very things they, the scientists, are unable to provide convincing proofs for. The unfortunate outcome of this is easily witnessed by one and all. Just remember the headlines over the last few decades...

The Hole in the Ozone layer — now seemingly fixed.
Global warming due to human activity, despite the fact that average world temperature has been dropping for more than a decade.
Eggs are bad for you—but now they're not!
Computers will provide humans with a 3 day working week: huh?
People are now living longer: no they're not. People are having less children, tilting the population age range.

And... this one is very ‘iffy’... constants are constant: light speed, gravity, etc.

But actually they fluctuate over time (if you study it all properly, you’ll see it too). They are actually measured to be different at different locations on earth too over time, but they are averaged out by institutes responsible for reporting them. The differences are very small and blamed on measuring errors. Are they? More importantly... are the constants really dead-on, precisely constant? Or is it more a case of where science-orientated formal bodies, responsible for reporting them faithfully, adjust the differences to a mean or average because... well... um... {blush}... everyone knows they are constant, don’t they? So, someone best make sure they stay that way!

A recent report which made me smile is related to Nov. 12, 2014, and the European Space Agency's Rosetta spacecraft and the Philae Lander.

They discovered carbon on the comet. Well, knock me down with a feather. As carbon is the soot of all stars/suns, and the 4th most abundant material in the known Universe, what else did they expect to find? Santa Claus’s grotto? (Excuse, please, my cynicism!). *Otherwise a great achievement you ‘guys’ at the ESA.*

* * * *

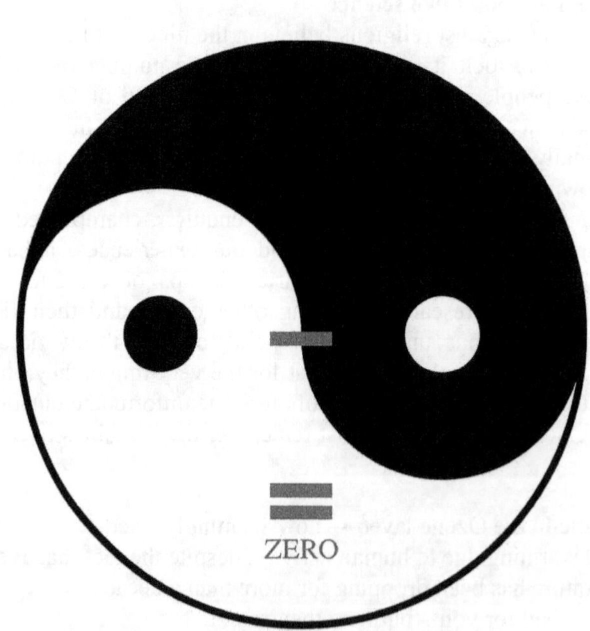

Why is there anything?

Chapter 6: Why is there anything?

This should be one of the hardest questions to answer, but, amazingly, it's the easiest. The answer is simple: the reason everything exists is because there is not nothing.

Well that's what logic dictates but it does not do it for me, nor I suspect, for you. It probably is an unanswerable question, but it doesn't mean we can't have a stab at it. We could start with the idea of the Big Bang, but much like the 'chicken and egg' enigma, the Big Bang itself is a difficult event to believe in if we simply take the notion that nothing at all ever existed prior to it. Most of the leading scientists who explore this quest (Stephen Hawking, for example) themselves do not subscribe to the idea of our universe/reality coming into existence from nothing. They believe the Big Bang was a result of 'something' already existing undergoing a change of some kind. Where this something is, or was, can be speculated as being both outside of *our* current reality/dimension and *our* time, since *our* time, and our perception of where we are now is within a place which started with the Big Bang itself.

This question is the most profound of all the big questions, because without a reasonable conjecture for the start of reality, everything is up for grabs—your mind, matter, your body, the stars... all might be a kind of illusion from the very start. It is possible we are just a dream inside a dreaming machine, constantly labelling and breaking down into labels, the artefacts which constitute the dream itself. But because this type of thinking and conceptualisation is unacceptable to the advantage of each person to experience a better dream, we would all reject such a notion: the more you take on the dream, and deal with its rudimentary laws and accept the illusion, the better odds you have of enjoying or advantaging yourself and your experience. Or so materialists have led us to believe.

Materialists are the dream believers. Without any real understanding of how reality manifested itself, they seek to adhere us to the dream. Science has not risen to be the lance to probe the veil, it has become the veil, evolving from mist into tempered steel, and quickly becoming the cloak itself that denies the lance a thrust! The current notion, which is more like a philosophy than anything else I've heard, is the concept of 'vacuum space'—also called, quantum vacuum fluctuation. This proposes a fluctuation took place in an ultimately finely balanced nothing! (Cough!} Pardon? Was Santa Claus or the tooth fairy involved, I wonder?

Let us go on our own journey and explore how reality might have been manufactured and for what reason. And let's begin with the materialist scientific view first to see how that one stands up to scrutiny. This is going to make your head ache, and this is why it is commonly ignored. I'll try and simplify, but if you are far more informed than my reductionist approach to very advanced abstract conjectures, you will already know, like me, the phantoms on which our comprehension of reality have been derived.

We have to go to back to the beginning, before our universe was here, and follow the events scientific study and investigation think happened.

The lead block or the nothingness?

Imagine with me for a moment a vast almost infinite block of dense lead. This metal is soft but wonderfully compact, heavy, self-joining. Imagine it is the only thing which exists... everywhere. I say this because when people try and picture the Big Bang, they usually think of a vast space filled with... well... nothing! And then this tiny spark goes *boosh* and a universe is born. But in our concept of the Big Bang, there is no space. So to get everyone nearer to the problem, it's better to consider a dense block of metal which itself is inert and set as an infinitely, all-pervading presence.

Got it? Right. Now for some unknown reason (which we will ignore for a moment only), a tiny expanding hole appears at its centre and expands, moving aside the lead. It carries on expanding, seemingly never stopping. Well, it hasn't for 13.5 billion plus years. This is the problem we face when imagining the Big Bang. Yes. You can replace the lead with nothingness if you wish, and it kind of makes it easier to accept, doesn't it? The problem comes now at this point. No matter whether your little hole of reality is expanding in lead or nothingness, or a sometimes fluctuating vacuum field/space—you, I, everything, is not within the expanding space, but on the thin surface seperating the hole from the lead, or the hole from the nothingness. The following simple diagram (I) below illustrates...

But even this might be wrong. We can only really map stuff onto the

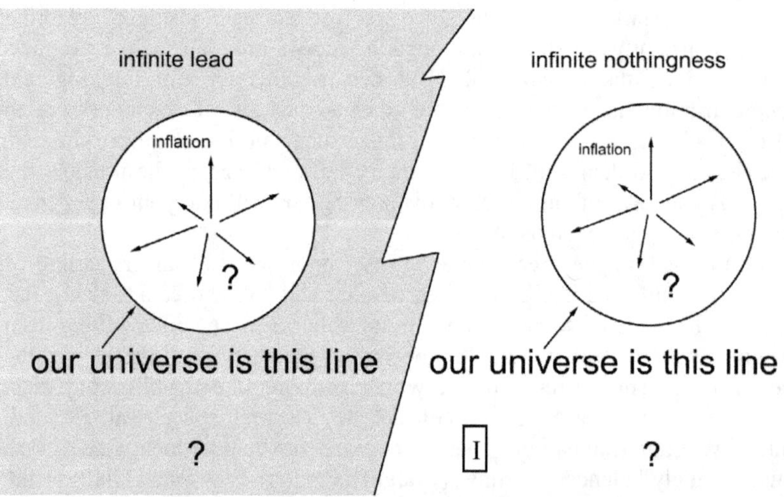

diagram where we know something measureable about it in a comprehensive way. If we do that, our thin line of reality where we know something about what's in it, looks like the variation on the opposite page (II).

The reason our line is almost non-existent is that everything on Earth, everything ever observed with all of our instruments, all normal matter out there in the cosmos, our reality—adds up to less than 5% of the Universe. We are only aware of, and have some understanding of 5% of reality!

The other 95% is up for grabs. Even if we get to know a little more about Dark Matter and Dark Energy, which is what most of reality is made from, we will still be left in a position of understanding absolutely nothing about what

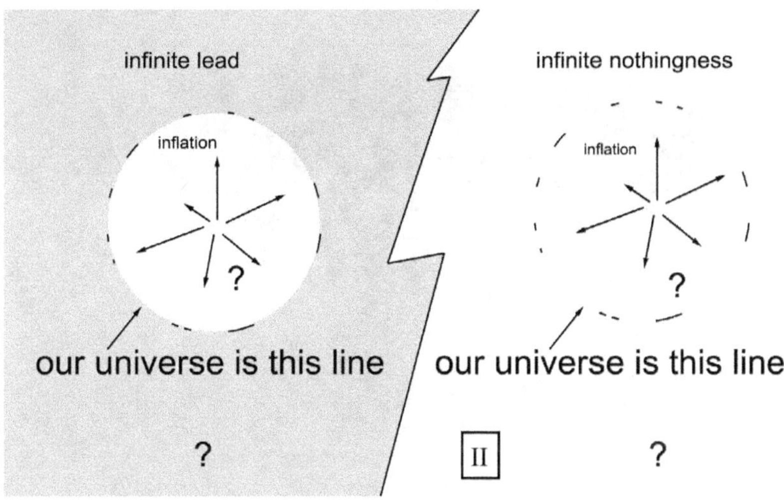

lies within the hole of expansion or outside the whole bubble. These spaces exist in a realm beyond any likely capacity to penetrate using a materialist approach to understanding everything. Materialism, our science, can only ever define more about that tiny thin line!

Everything on that thin line, seen or not seen, understood a little or not understood at all, everything which constitutes the concept of what we term as matter, and all the unknown stuff, is growing further apart at an ever increasing rate. Reality as an abstract and as a physical reality is in many respects growing fainter!

The NASA diagram on the next page illustrates this.

Don't kid us along

The simple truth is this: ordinary humans have a problem imagining a nothingness. It is just not in our collective psyche. The Romans had an issue with it too. They never invented a zero symbol to represent it. Infinity, eternity, zilch—they are all metaphoric concepts lacking any material and real presentations. We don't 'get' them. The tool of mathematics accepts these limits and mathematicians devise clever ways to work with them. Science is now 'preaching' there never was nothing after all. There was always 'something' even if that something was the nothingness of quantum vacuum space. Quite frankly, I can't see a lot of difference between this idea and the one of 'God moving across the face of the deep, and saying..'

But. I'm sure you are familiar with that one too?

When you look at my simplistic diagrams prior to the NASA expansion diagram, you may think the universal expansion is being caused by some kind of force exerted from inside the hole of the diagrammatic circle of reality. We imagine balloons inflating and it's easy to ascribe that notion to the larger and more mysterious issue of an expanding universe, right? Well, that's what I'm guilty of too. But it's not like that. We don't know what is causing expansion. The thin line is made up of visible light matter (5%) and of invisible Dark Matter and the invisible Dark Energy—95%. The larger portion of reality, the

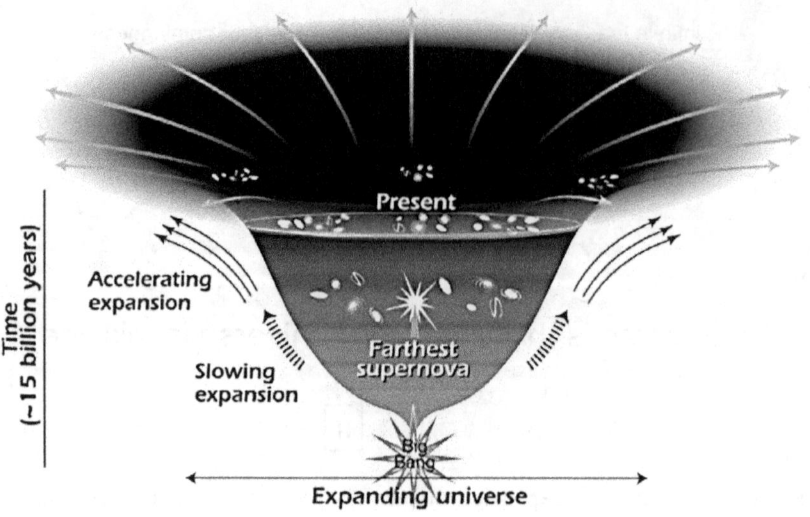

This diagram reveals changes in the rate of expansion since the universe's birth 15 billion years ago. The more shallow the curve, the faster the rate of expansion. The curve changes noticeably about 7.5 billion years ago, when objects in the universe began flying apart at a faster rate. Astronomers theorize that the faster expansion rate is due to a mysterious, dark force that is pushing galaxies apart.

dark matter/dark energy may itself be the driving force, or not. It might be that bits of reality, like most of it, possesses a trait that pushes what we consider as normal reality, the normal matter that we live on and are composed of, further and further apart!

My point here is this. We know so little about what reality actually is. Yes, we know something about physical real stuff, like metal, molecules, crystals, uranium, suns, planets, energy (white energy, white matter, energetic material which activate our senses and the physical instruments created to extend our senses), but what our journey of scientific endeavour has shown us is our ignorance of the magnitude of the problem. The journey of materialistic science is reaching a problematic limitation. It cannot penetrate what cannot be seen or detected by physical sensory constructs alone. The material scientists are in trouble!

To their aid, like the Calvary, comes the scientific mathematics too obscure to all but the most 'gifted' of minds to understand, and with it—the mathematicians, the people who define reality as relationships between numbers and more numbers which map cause and effect, but describe nothing of value which can be imagined or accepted in most human minds. And this includes the mathematicians themselves. Their maths cannot create true comprehension for them either. It only exposes working methods enabling humanity to exploit matter and mould it into useful aids for our survival, or create powerful killing machines and methods to end our journey.

It is as though a flower is presented in colour, scent, and form, and we would wish to understand all a flower is, but instead use a single eye to behold it, and a single sentence to describe it. Science, as a discipline, writes the sentence, and no one reads what else could be written. A poet may abstract the flower, a writer may paint it with words, an artist with images and form, a chemist with reductionist certainty into its essential oils, a botanist to its genetics, a mathematician to a set of numbers which describe its physical space and its Fibonacci number coincidence in its petal formation; someone else might see it in all of its invocations but could never reduce it to one or another in preference. And that person's experience will be ignored by the collective route of scientific investigation. All that will survive will be those elements and notions which fit into an entrapping and already-constructed mould.

And yet we all experience a flower. Our minds can quickly accept both metaphoric references to birth, emergence, beauty, age, and death. We can all see the flower is part of life. Bees visit the flower. We consume the honey the bees make. We smell the scent. We recognise the brilliance of colour produced from brown and dull earth. A human mind can understand the flower as an abstract, a metaphor, and as a physical dynamic mechanism. This, I suggest, is a full understanding of what a flower is. A human life is enriched by the novelty of that experience.

So, to even to try to begin to understand the thin line of reality—known reality, in any kind of pertinent way, the question is not about what reality is, but more about what if offers you—an emergent part of it. To describe it, to reduce it to anything a human being is unable to understand, adds nothing to the experience of being alive... unless it is relevant to remaining alive (materialistically?) and enjoying the experience of that (metaphysically).

It seems to me that we live in the light matter and not the dark stuff. I doubt from what I know, which is as little as all humanity knows, that in the 95% of the unknown universe, there is any life—there, in the Dark Matter, and that life itself is a product of the things we can detect readily because we are part of that stuff and are made in such a way to see only what we are made from. I see light because light sustains me. I am made of carbon because my sun pours it out by the ton every second. I live on earth because this clump of 'light-matter' settled here and draws energy from that star close by.

The reason why there is something rather than nothing is because I am part of the something I can see and draw experience from it as I go through my cycle of birth, flowering, demise and death. My material is not wasted. All I am physically, will be reused to build new experiences for all living things as long as they remain in the light. My mind may reside in this organic host or just pass through it. My mind is not physical even though it persists through the exploitation of physical matter and its processes. A physical process cannot understand anything, only a mind can. A mind is the only construction the universe, my universe, my reality, can seem to manifest upon physical light-orientated matter to experience the extraordinary sensation of 'being' instead of not 'being'. It has somehow managed through its 5% clusters, spread out within a mysterious and ever present majority of something else expanding within our reality space, an incredible thing. It has created experience from ordered chaos.

It may be limited, a few years within eons. It may be fleeting. One thing it isn't though, is a set of numbers, or a material-only event. It is a curiosity. It is a stunning and unexpected novelty to be born of starlight and dust, to suck in gas formed from the fall out of creation, to find myself here on a lump of that original material, now less energetic, whizzing around a nuclear fusion furnace, in the arm of a billion other such systems, all running like hell away from the dark spaces of reality, of which we have no knowledge.

I suspect, as I am here and able to perceive this event, there is a purpose to it, despite the fact I am not born knowing that purpose. I can imagine and conjecture all kinds of reasons why I am able to experience what I do. I cannot accept my life or yours to be for no reason at all. This is not based on hope, but reason: it would be easier for the universe just to unfold as a mechanistic process and be whatever it will ultimately become, than it would be to somehow construct replicating structures, struggling against entropy, and becoming witnesses not merely observing, but involved with the even itself!

The reason why there is something rather than nothing is because an event is going on. It is an event I am part of. So are you. That event doesn't just affect me, you, the ants, the flowers, the planets, the suns, it affects all of reality. It's a biggie. It's an event so big that a single lifetime is insignificant when weighed up to the absolute outcome of all reality. It may be that a something which was quite 'happy' to be as it was before the event started, somehow got nudged into a different state of being, and now desperately reaches within itself to make sense of it all and work out a way (if possible) to return to its desired and natural state. It might be an event that has been going on in some rhythmic repetitive and permanent form forever, and that is why something can persist rather than nothing.

In all of the things I have seen in life, one thing is true. Action and activity happen most fervently at the boundaries of incompatible systems. Be it ideas, forces, or matter—where tension and polarisation exist most, there we find the greatest activity. There may be echoes in human existence and experience mirroring the grander play of a universe in chaos and unresolved tension!

Maybe the reason there is something rather than nothing is something out there, something in here, something about reality cannot fully recombine into a whole from its many parts, and yet it seeks to. The event we witness and are part of, might well be a continuous failed attempt to re-combine something which cannot be brought together as a whole. Maybe there is always something because the absolute nature of things dictates there can never be nothing: an imbalance, no matter how small, a polarisation, or an unknown, irreconcilable 'difference' inhibits the sum of all things from being brought back to a totality of zero. A tiny something—a fraction which defeats the summing to nought— is always present and thus gives rise to a reality which will always exist.

But this is a philosophical approach, yes?

Strange as it may seen, it is also the materialistic scientific approach too. Different tools: same conclusions! Let me try and explain the leading scientific minds rationale as to how a universe started from nothing.

The idea is that 'nothing' is the stable state of a non-system. If you wish, you can imagine it to be an abstract where all potential infinite energy and mass is there, in the nothingness, but only as potential. The nothing exists because there is no tension, no disruption of the nothing state. However, the merest, tiniest introduction of anything, the smallest fluctuation in a quanta of energy, would lead to rapid instability, and cause an event like the Big Bang—where an extraordinary and immeasurable release of energy and matter takes place. The energy and matter created would have consisted of positive and negative forms, matter and anti-matter. These opposite forms would have collided and annihilated each other. But, against the odds, they didn't. What appears to be left, and what our universe and reality appears to be, is the legacy of all the positive matter and energy which had no oppositely equal and cancelling counterpart.

Although 'they' don't say it, I will: the universe appears not to have come from nothing; it appears it came from a tiny imbalance in the polarisation potential of the nothing state. For all we know, the nothing-state may only have existed for the merest billionth of a nanosecond—possibly a state which itself was a legacy of a collapsing previous reality. The universe may be more like a pump, or a set of bellows, or the spark across a spark plug gap. The universe may one day stop inflating and just simply reverse and collapse. All scientific theories regarding the creation of the universe and its likely outcome are only theories and highly speculative! All of them have profound and unsolvable problems attached to their credibility.

Theories exist that state a vacuum can fluctuate at the quantum level and give rise to the emission of particle pairs. Er... what vacuum are we referring to exactly? A vacuum is a component of a universe. Without a universe, there is no vacuum, only the dense infinite lead of nothingness.

Theories exist which suggest our universe was created as a by product of two or more higher dimensional 'Branes' colliding in that dimension. The term 'Branes' arises from the foreshortening of the word membrane. Er... but then that pre-supposes something did exist, and goes on existing, outside of our universe and begs the ever circular question of how did that get created?

People who believe in a god or who have strong religious beliefs quite rightly pick substantial and credible holes in all the scientific theories regarding creation. By showing that scientific conjecture regarding creation is just a belief system itself, they can propose their belief system—the one of thinking a sentient creator made the universe—is just as believable as the scientific one. And they are right. Both are the same: both are belief systems.

You can invent your own if you wish. We could be living inside a computer. We could be the dream of a whale. We could be our own dream—an escape from an eternal sleep as abstract forms in an abstract space which has no foundation in any physicality at all. The exposing of sub-atomic particle behaviour in a quantum world, and the labelling of forces at work in our universe acting on those particles both at the microscopic and macroscopic scale, is no more a proof of creation than are the words, sentences and paragraphs discovered in Genesis in the bible.

It seems you pick a side and believe in that. This itself is a problem which science has failed to avoid in its evolution. They may as well include the idea of a creator in their conjectures regarding creation as it merely provides another choice of many ideas which can never be possible to prove. Unfortunately, the struggle in the early days of scientific reasoning, to take hold of mass public acceptance, meant a collision between powerful forces here on earth—namely religious dogma and the power religion had over billions of lives, and the emergence of testable reason. It is a sad indictment that many followers of science have become the most militant atheists and the most religious zealots have become stone deaf to scientific reasoning.

It's sad because the many people in each corner of the debate over creation are in fact the few. Most human beings can believe in two things at the same time. They can imagine the Big Bang as coming about through some kind of natural consequence, and they can still believe in the possibility of a god or gods being involved with the circumstance which led to the natural world starting our universe and reality running. The human mind is capable of perceiving and accepting what others argue are mutually exclusive ideas. Clearly, they are not. At least fifty per cent of scientists are open minded about the existence of a purpose to life and the idea that a sentient mind or minds was/were involved in our universe's creation. They are just not allowed to say so. They belong to an order presided over by the new priests, predominantly esteemed men, with reputations to preserve, and a belief in scientific thinking being superior to other modes of human thought. The 'troops' must keep their mouths shut or lose their jobs and credibility because the people above them in their hierarchical system will make sure that happens.

Anyway, I'm drifting off the subject so let's just summarise.

The reason there is something rather than nothing is because there has never been nothing. Something has always existed. It may not be a universe. It may not even be matter or energy as we understand it to be. Matter and Energy are facets of our universe. They don't have to be facets of the whole of whatever it is our universe is part of. The whole thing, the ultimate something, does not have to be space or anything physical at all. It can actually be an abstract. Your mind is an abstract. It doesn't exist as a thing itself. It is a manifestation of the properties of matter and energy interacting together in a unique formation, either inside a brain (the materialistic view), or through exploiting the brain as a kind of tuning device.

Without mind being in existence, there would be no question to ask about why there is something rather than nothing. Possibly, without mind, there may be no universe, as there would be no way to prove it existed or not. We might all be sharing the same dream and the universe is our dream. No one can prove anything on these big issues. All can conjecture, all can pick a conjecture that's right for them. Or, all can keep their minds open and not decide on any answer. The choice is theirs alone... as is yours.

One thing I will say, the chances of you being here is just so vastly against the odds, you should not be alive at all. *In fact, the logistical odds from materialistic thinking dictates it is impossible for you to be here!* The odds are so very bad. Even science has entered the area of philosophy to try to explain it. But they disguise it. Maybe we should use their ideas to explain what otherwise might be referred to by deeply religious people as ... well...

...a miracle!

* * * *

The Raising of Lazarus, (c. 1410) from the Très Riches Heures du Duc de Berry, Musée Condé, France. From wiki. Photos or art is public domain

Chapter 7: A miracle?

Scientific theories dictate that Biodiversity is the result of 3.7 billion years of evolution. The origin of life has not been definitely established by science, however some evidence suggests that life may already have been well-established only a few hundred million years after the formation of the Earth. Life only consisted of archaea, bacteria, protozoans and similar single-celled organisms until approximately 600 million years ago.

Evolution seems to be the cause then. Now, I have to admit that when I look down my microscope at a single bacterium, and this is what I see...

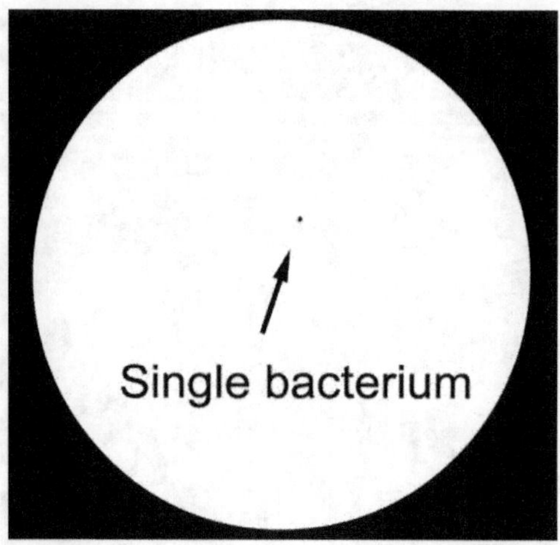

...and given that a single bacterium is made up from 139 kbp to 13,000 kbp (Kilo Base Pairs: building blocks of the DNA double helix), and that a human being consists of 3 billion base pairs organized into 23 paired chromosomes, it seems highly unlikely I evolved from that little dot, despite the possible 600 million years of mutation which evolution theory insists occurred to make it happen since first bacteria to living me.

It appears even more profound to me that the little dot, with quite a limited program-like code formed of base pairs, can also simultaneously, through repetitive mutation and iterative reproduction, re-pattern itself to produce over 8 billion to a possible 80 billion individual different species!

In fact, the whole idea appears non-evidential, and quite absurd. It also seems absurd to suggest those first forms of life came into existence through a mucky pool of the right chemicals swimming around looking to combine into more complex, self-sustaining forms.

And when you look inside even the smallest of living structures, what do you observe? Organisation, and cooperation between the organised structures and processes. (See diagram on the opposite page of a bacterium—[singular]).

Yet again, an 'intent' to self organise, and to cohere, appears to drive and sustain this tiny packet of life. Those characteristics did not create themselves,

From Wiki (creative commons licence).
Original creator. Mariana Ruiz Villarreal

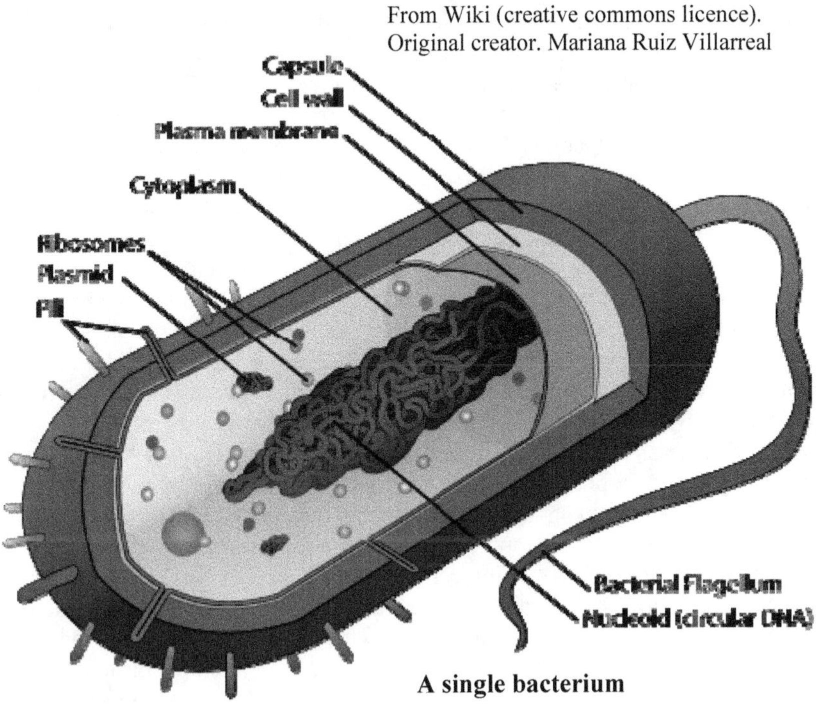

A single bacterium

but if they did, it is because other prevalent and abstract properties already exist in the very atoms of the material components to make that happen; the atoms imbued the structure with these traits. Maybe smaller structures too—sub-atomic particles, and their primal quantum activity assisted.

Let's just say evolution in all the aspects of its theory is true. Somewhere on primordial planet earth, amino acids, electrical discharges, gases, and an invisible but clearly inherent bias in each atom of each molecule and compound to merge with whatever it can to make something new, or to seek improved stability, created the first organic structures came into existence on planet Earth. (*And that takes a lot of believing for me, because belief is all we have. There is as yet, no scientific proof of the way life has apparently emerged from non-living things*).

Probability of self emergence of life through random chance

The argument from unlikely-probability that life could not form by natural processes, but must have been created or biased to happen is almost irrefutable, but the issue is often ignored by pure evolutionists; although many acknowledge it is a strong argument for the existence of an intelligent creator intent on making life happen. The probability of the chance formation of a hypothetical functional 'simple' cell, given all the ingredients, is acknowledged to be worse than 1×10^{57800}. To put this in perspective, there are about 10^{80} electrons in the universe. Even if every electron in our universe were another universe the same size as ours that would 'only' amount to

10^{160} electrons. These numbers defy our ability to comprehend how very impossible life is! Fred Hoyle, British mathematician and astronomer, used analogies to try to convey the immensity of the problem. For example, Hoyle said the probability of the formation of just one of the many proteins on which life depends is comparable to that of the solar system packed full of blind people randomly shuffling Rubik's cubes all arriving at the solution at the same time; and this is the chance of getting only one of the 400 or more proteins of the hypothetical minimum cell proposed by the evolutionists ('simple' bacteria have about 2,000 proteins and are incredibly complex). The order in the proteins and DNA of living things is independent of the properties of the chemicals of which they consist—unlike a snow crystal where the structure results from the properties of the water molecule. The order in living things parallels that in printed books, where the information is not contained in the ink, or even in the letters, but in the complex arrangement of letters which make up words, words which make up sentences, sentences which make up paragraphs, paragraphs which make up chapters, and chapters which make up books. These components of written language respectively parallel the nucleic acid bases, codons, genes, operons, chromosomes and genomes, which make up the genetic programs of living cells.

And if that 1 in 10^{57800} chance happened, a miracle by itself, it's even more extraordinary to presume the notion that the first living structure on Earth would already be imbued with a mechanism allowing it almost instantaneously (a few hundred million years is nothing compared to the age of the universe) to start transforming into an enormous, mind-boggling range of vastly more complex living forms.

There are a known 8 million different species on earth and a view in existence by experts that we may have uncovered only 10% of all species, it would seem evolution has to produce on average (if it were working at a linear steady rate), one new species every forty-three years. Of course, we are only living with the surviving mutations. Since mutation is random, and most mutations would probably cause a quickly-dead species, we could make an assumption that evolution makes ten badly formed mutations that die out almost immediately in the first form to every one species that kind of works out. I suspect the ratio would be vastly different to this in magnitude: try rolling three dices to get three sixes!

But this suggests evolution has been producing not one new species every forty-three years, but ten, or one species every four years. They may not have been made linearly like this, of course, but it puts things into perspective.

So, presuming all life evolved from just one muddy pool (soup) through an extraordinary lucky mix of just the right chemicals, and ambient surrounding conditions, the ingredients must have been incredibly lucky to have found themselves in that one place at that single moment to go on for 600 million years, churning out all these different variants of their original coherence. Or did luck have a guiding hand? *See my graphic at the top of facing page.*

Making a thing which works through mutual cooperation of its parts

Matter and Energy play with their possible mixes like blind throws of dice.

10x10x10x10x10x10x10x10x10x10x10x10x10x10x10x10x10x10x10x10x10x10x10x10x
10x10x10x10x10x10x10x10x10x10x10x10x10x10x10x10x10x10x10x10x10x10x10x10x
10x10x10x10x10x10x10x10x10x10x10x10x10x10x10x10x10x10x10x10x10x10x10x10x
10x10x10x10x10x10x10x10x10x10x10x10x10x10x10x10x10x10x10x10x10x10x10x10x
10x10x10x10x10x10x10x10x10x10x10x10x10x10x10x10x10x10x10x10x10x10x10x10x
10x10x10x10x10x10x10x10x10x10x10x10x10x10x10x10x10x10x10x10x10x10x10x10x
and on and on and on times 10 another 57,656 times!

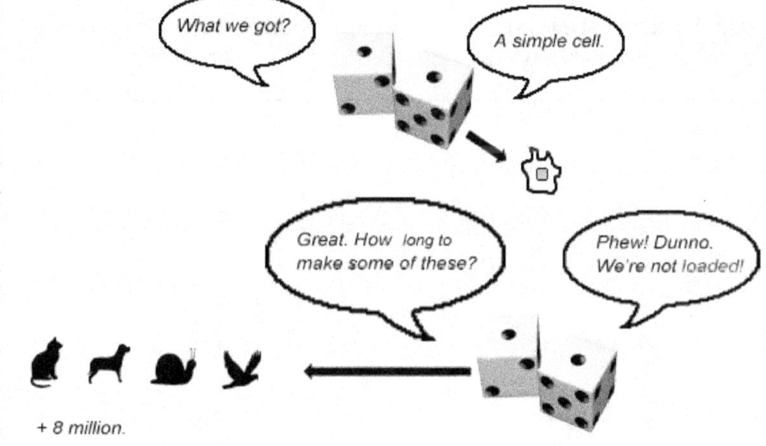

The probability of randomly creating 8 million species of life if there is no bias in the process.

I could cut a load of fine metal into tiny cogs, springs, levers, ratchets, and spindles, chuck the lot into a slim metal case, stick a dial on it and a sliver of glass, but no matter how long I wait, I won't ever be able to use it to display the time. The watch doesn't work unless all the bits are in the right place and the complete unit is organised such that all its parts co-operate with each other to carry out their functions for the good of the whole. The good of the whole in this example is 'display the time'. All living things are constructed from tiny organic components. Something organises life into discrete packages, and each package is effective at performing the function it seems designed for. Something may also be making life as part of the function of the whole. Like the parts in a watch to tell the time, the entire universe and the life within it may also be working in cooperation for a purpose unknown to us.

In the natural world where there is no apparent designer, science ascribes the organisational role to the existence of several universal laws and processes interacting, with a slim opportunity for giving rise to life as part of these processes. But science offers no complete solution for why, or a coherent and acceptable reason for the emergence of life.

The Secret Of Life?

A few complex chemicals make life. But unfortunately, living forms need a constant supply of their constituent parts to remain stable (since they all die) to at least live long enough to replicate themselves in slightly variant forms. This means they need a food chain and an energy supply even to fuel the lowest forms. Sometimes, the simpler living forms at the bottom of the chain, which have nothing to predate or eat, have to magically make their physical needs from something not living to continue to exist.

Enter the story—Diatoms, Plankton, Phytoplankton, Algae

We can't eat rocks but we need minerals and inorganic matter to maintain our

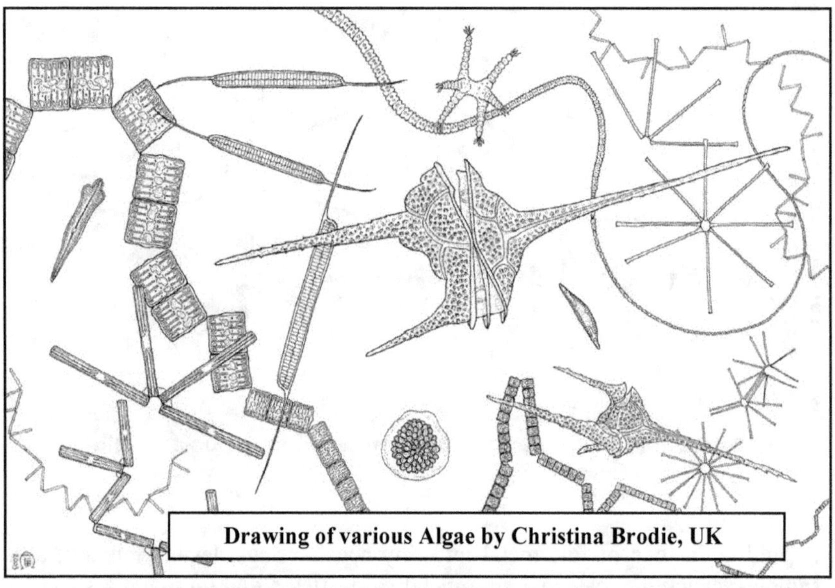

Drawing of various Algae by Christina Brodie, UK

structures. Diatoms just need rocks (minerals), and sunlight... oh... and vitamin B too. Living things on earth also need oxygen. Irrespective of this, it doesn't matter if the gas sustaining them is oxygen or carbon dioxide, they all need water—H^2O. Everything living depends on water and energy.

Algae perform the extraordinary task of converting direct sunlight into physical matter: chemicals! Photosynthesis is a process used by plants and other organisms to convert light energy, normally from the sun, into chemical energy that can be used to fuel organisms' activities. Carbohydrates such as sugars, are synthesized from carbon dioxide and water (hence the name photosynthesis. Oxygen is also released, mostly as a waste product. Most plants, most algae, and cyanobacteria perform the process of photosynthesis, and are called photoautotrophs. Photosynthesis maintains atmospheric oxygen levels and supplies all of the organic compounds and most of the energy necessary for all life on Earth. And of course, these living forms exercising photosynthesis are eaten and thereby maintain the process of re-distributing important nutrients up through the food chain.

If I may generalise for a moment, at least two fundamental and critical processes are required for life as we know it to exist. Not photosynthesis or the presence of oxygen. I'm leaving these out along with many other interdependencies which also must be maintained in balance. They are not what you think, and they lead to exotic clues that life and our reality may not be what it appears to be at first glance. We'll get to those in a moment. But first, if you look at some of the microscopic forms on the previous page, you will notice they appear to share something in their construction. A kind of similar pattern emerges, complex and compelling! This patterning is something they seem to share. What do complex patterns tell us?

Patterns are formed by influences

The closer you look through a microscope, the more things you observe, the more aware you become that similar patterns are emerging from a set of 'influences' which science has deemed the result of mathematical processes—the same underlying behaviours which produce fractals.

Observe the photograph of a sea urchin below. Next to it is a Fractal

Image of Sea Urchin (Wiki creative commons licence). Fractal public domain by Google.

Image generated by a computer resolving a set of mathematical formulae and producing a visual interpretation of the result. Any similarities?

You have to think abut this quite carefully. Using a few mathematical formulae and a computer to produce visual geometrical solutions or interpretations of the interactions and relationships between a few variables, we are able to generate shapes and forms similar to the living and non-living structures we see around us; structures formed by invisible influences inherent in the properties of matter itself in our universe. We also have to think about what we mean by mathematical formulae.

To me, this is maths: $2 + 2 = 4$ or where $A = 2$ and $B = 4$, $A + B = 6$. It is a way of showing numerically the way two known quantities can be ascribed a total quantity representation - a number. Sometimes, we know a lot about a set of unknowns and their relationship to each other, for example we have Ohm's law: I = V/R where I= the current, V = voltage, and R=resistance in the circuit. And if we don't know one of the variable's value, because we know the relationship, we can ascertain the unknown quantity. Mathematics remains to be this when quite complex relationships between different variables are involved.

I believe it is important to try and distinguish the essential difference between these two statements: add A and B to make C, which is profoundly different from A + B = C.

The first is an instruction. The latter an observation! Computer code is a set of instructions which uses mathematics as part of the process for resolving an outcome as a product of those instructions. Invariably, a computer program has an aim, or is exploring ('testing') a method to find a way (a solution) to solve a problem. In all the things one tends to examine within the microscopical world and the cosmos as a whole, it is clear and transparent to many intelligent minds that the universe and all it contains, appear to be the result of iterated code generating new code and all physical entities are driven and manufactured from that code.

fractal shape form of a Romanesco broccoli
Jon Sullivan
(Wiki commons licence)

Image (top) is a fractal rendering which produces an image similar to the vegetable broccoli when an instruction set is applied to use the formulae set and compute an image of a few variables' relationship to each other. The photograph on the facing page is of water vapour (moisture) freezing and fracturing on a glass window pane. The latter process appears to be the result of an instruction set covertly mapped onto the physical material called water

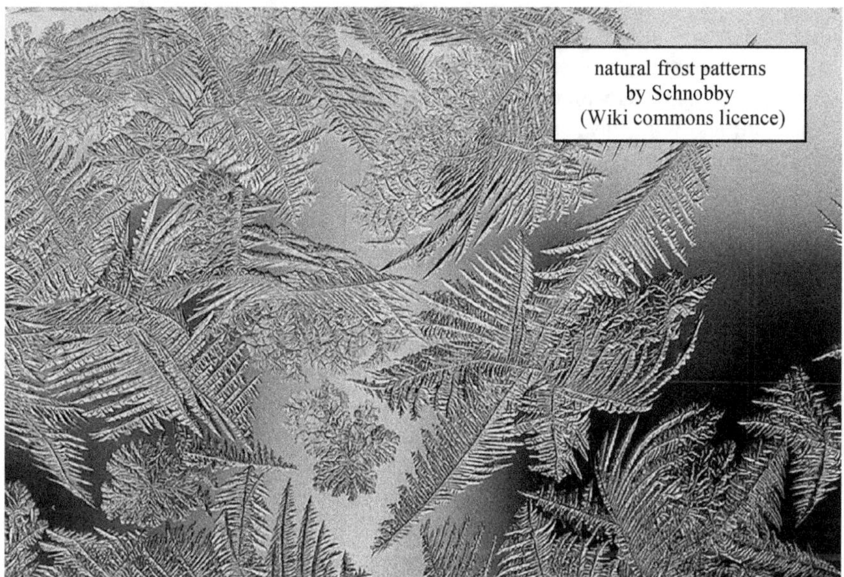
natural frost patterns
by Schnobby
(Wiki commons licence)

and its relationship with heat energy, or the lack of it. The result is a computation which is using mathematics but it is not the result of passive relationships. It is the result of dynamic obedience to a set of instructions.

We can split hairs about our labels of mathematics and program code, but mathematical formulae not applied to anything remains just that—nothing. To apply it, requires a set of instructions.

By deduction and comparison with our very new tools of computers, logic, computer modelling, and their application (something which Newtonian Scientists, Philosophical Mentors, and Spiritual Thinkers of the past did not possess or utilise in their thinking and work), we can look at our reality completely differently from anyone who ever lived before. We have fresh insights and new ways at looking at our universe.

We live inside a computer programmed Universe?

You can probably see where I am going with this. I am raising the question that we may be living inside a computer-styled and programmed universe. Which then starts to raise questions about 'intent' and whether or not such an entity could possibly create itself? I am a mere enthusiast microscopist, not formally qualified in any formal topic of education, and so it would seem very audacious for me to suggest such a thing. Don't you agree? Ok. Let's not take my word for such an idea, let's go to someone much respected in the scientific community and see what he has to say.

Enter Stephen Wolfram

From 1992 to 2002, Wolfram worked on his controversial book, A New Kind of Science, which presents an empirical study of very simple computational systems. Additionally, it argues that for fundamental reasons these types of systems, rather than traditional mathematics, are needed to model and understand complexity in nature. Wolfram's conclusion is that the universe is digital in its nature, and runs on fundamental laws which can be described as simple programs. He predicts that a realization of this within the scientific communities will have a major and revolutionary influence on physics, chemistry and biology and the majority of the scientific areas in general, which is the reason for the book's title.

This scientist has explored more than most of us put together including the areas of Quantum Physics and Particle Physics. He is cited by over 30,000 publications. I am not saying he says someone or something made our universe, he isn't. I am saying our universe behaves like a computer, and he says that too. What I am saying more though, is whenever I have built a computer, it doesn't seem likely one could build itself. More.. it seems unthinkable it could built itself through chaotic and random experiment and by keeping only that which works, and discarding the rest. More-over, any such computer or the parts it’s made from, needs to have an intent to construct itself to begin that task. Where would such an intention originate?

Back to Fractals, Patterns, and Influences

I mentioned patterns are formed by influences, at least those manifestations we recognise as being patterns. Here are two visual patterns I produced at random in photoshop.

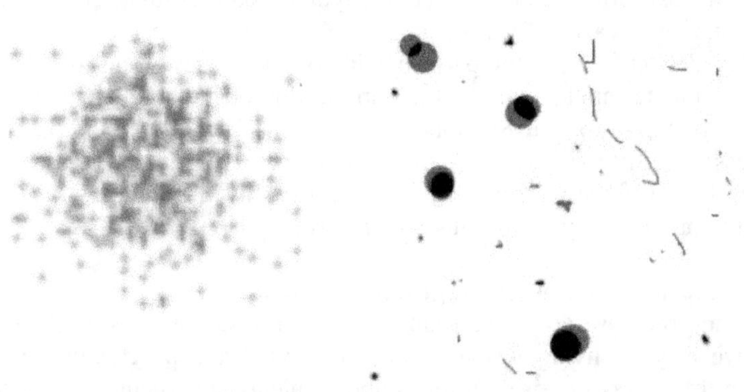

I hope you will agree with me that a few blobs of ink on white paper, which is not repeated within itself or in another image, does not seem to form what we recognise as a pattern. It's almost intuitive, isn't it?

Whatever influenced my hand (my thoughts) on the first image was not the same on my second image. But this is a common aim between the creation of these two images. I aimed not to repeat a similar looking set of blobs and I intended not to create images with any recognisable symmetry.

So... what about the image set on the facing page (1—4)? Do you see anything obvious there?

What you are observing are still frames from a computer animation of a fractal being computed of a Rössler system, a system of three non-linear ordinary differential equations originally studied by Otto Rössler. These

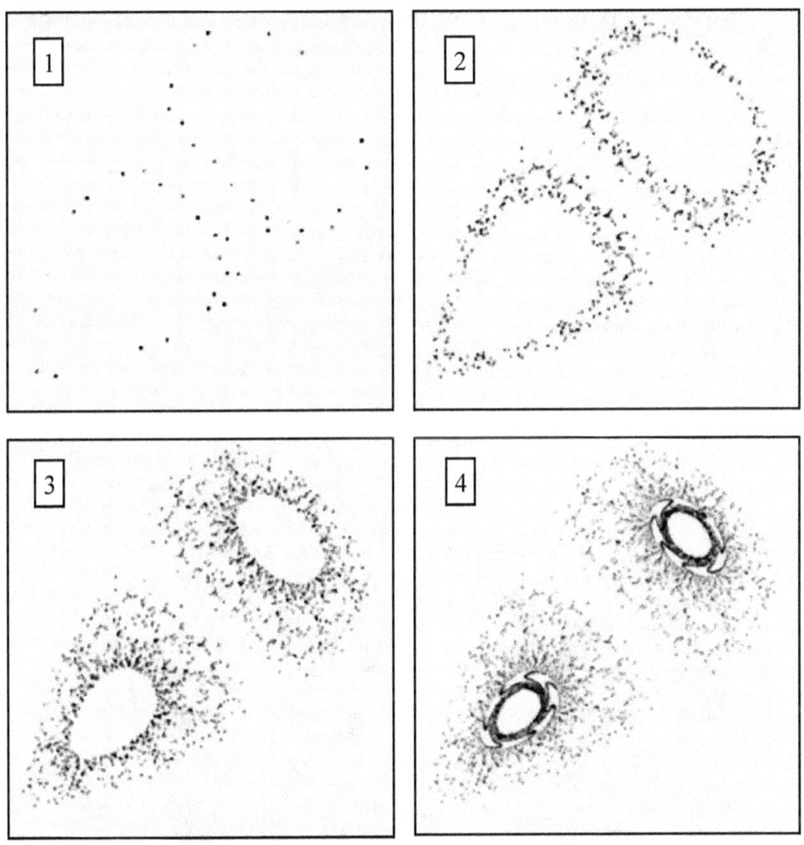

Generated with Fractal Growth Generator for ImageJ; software and animation image by A. Karperien. (wiki creative commons licence).
Full animation online here:
http://www.microscopy-uk.org.uk/mag/imgaug13/karperien_strange_attractor_200.gif

differential equations define a continuous-time dynamic system that exhibits chaotic dynamics associated with the fractal properties of the attractor.

An Attractor? Highly exotic! What's that? Fundamentally, it is a bias operating on a system and influencing it in a unique way.

The world of fractals may offer more insights regarding the behaviour of real processes in a real universe. The only method strange attractors use is changing numbers using formulae. This turns out to be very useful in studying nature, where we look at the ways things like population, weather, and chemical reactions change. Scientists found many fractal patterns in these natural changes. In fact, strange attractors like the Rossler Attractor and the Lorenz Attractor were discovered while studying natural, not mathematical phenomena.

Now look at the image on the next page. In this construction, an orbit within the attractor follows an outward spiral close to the plane around an unstable fixed point. Once the graph spirals out enough, a second fixed point influences the graph, causing a rise and twist in the other dimension. In the time domain, it becomes apparent that although each variable is oscillating

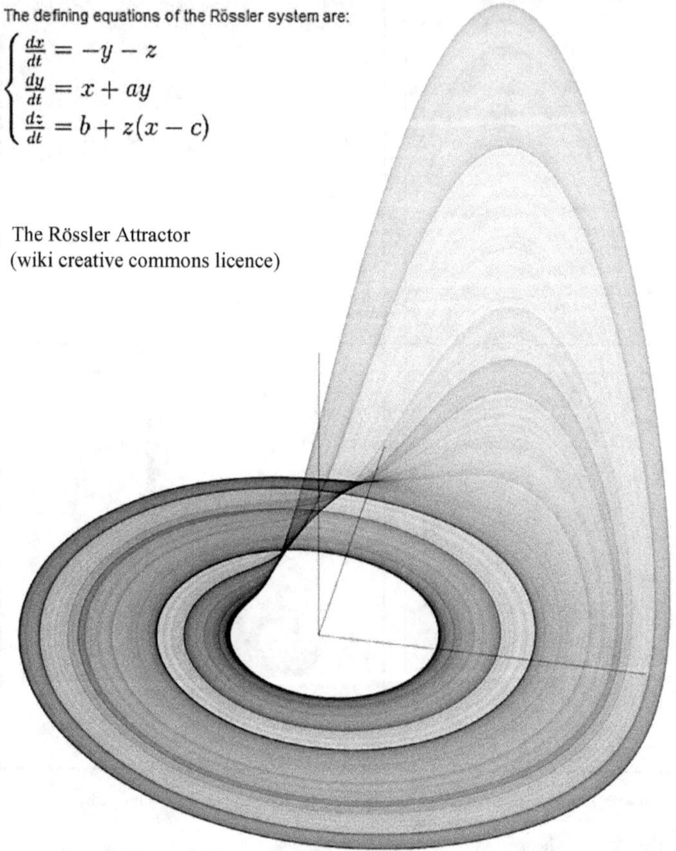
The defining equations of the Rössler system are:
$$\begin{cases} \frac{dx}{dt} = -y - z \\ \frac{dy}{dt} = x + ay \\ \frac{dz}{dt} = b + z(x - c) \end{cases}$$

The Rössler Attractor
(wiki creative commons licence)

within a fixed range of values, the oscillations are chaotic.

This attractor has some similarities to the Lorenz attractor, but is simpler and has only one manifold. Otto Rössler designed the Rössler attractor in 1976, but the original theoretical equations were later found to be useful in modelling equilibrium in chemical reactions.

On the facing page, an attractor is set towards which a variable, moving according to the dictates of a dynamical system, evolves over time. That is, points that get close enough to the attractor remain close even if slightly disturbed. The evolving variable may be represented algebraically as an n-dimensional vector. The attractor is a region in n-dimensional space. In physical systems, the n dimensions may be, for example, two or three positional coordinates for each of one or more physical entities; in economic systems, they may be separate variables such as the inflation rate and the

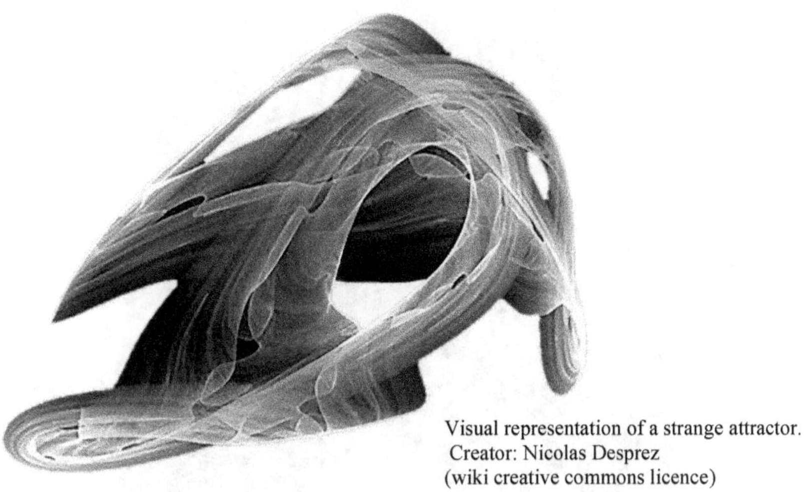

Visual representation of a strange attractor.
Creator: Nicolas Desprez
(wiki creative commons licence)

unemployment rate.

Now... do you recall the rolling dice graphic on a previous page? I was trying to indicate that life can only come about the way it has due to a predisposed bias in the chance system... an attractor? The argument is not a case of 'Does our reality have biasing attractors and repellers like the attractors in mathematical systems visualised by computer generated fractals?' Has life itself been brought into being from a hidden attractor creating a bias in atomic and sub-atomic systems and particles?

The argument is not a case of a Cinderella scenario for a universe which can support life: it is a scenario in which a clear-to-see bias exists to create life! The universe wanted you to be made!

If a bias exists in the underlying nature of reality for intelligence to emerge, it suggests that bias and potential for emergent intelligence, is created for a reason or for a purpose. Whatever it is, religious faith systems have tried to answer and modern science faith systems have remained deaf to.

Most human beings *believe* there *is* a purpose to life. Perhaps they, *we*, are smart enough to instinctively acknowledge that the experience of all the things we suffer and bear, the emotional drama of being, and the fact a universe somehow has brought us into existence, despite the enormity of its size and awesome energy—that all this would be such a waste of extraordinary and unlikely success, if there was no point to it at all.

* * * *

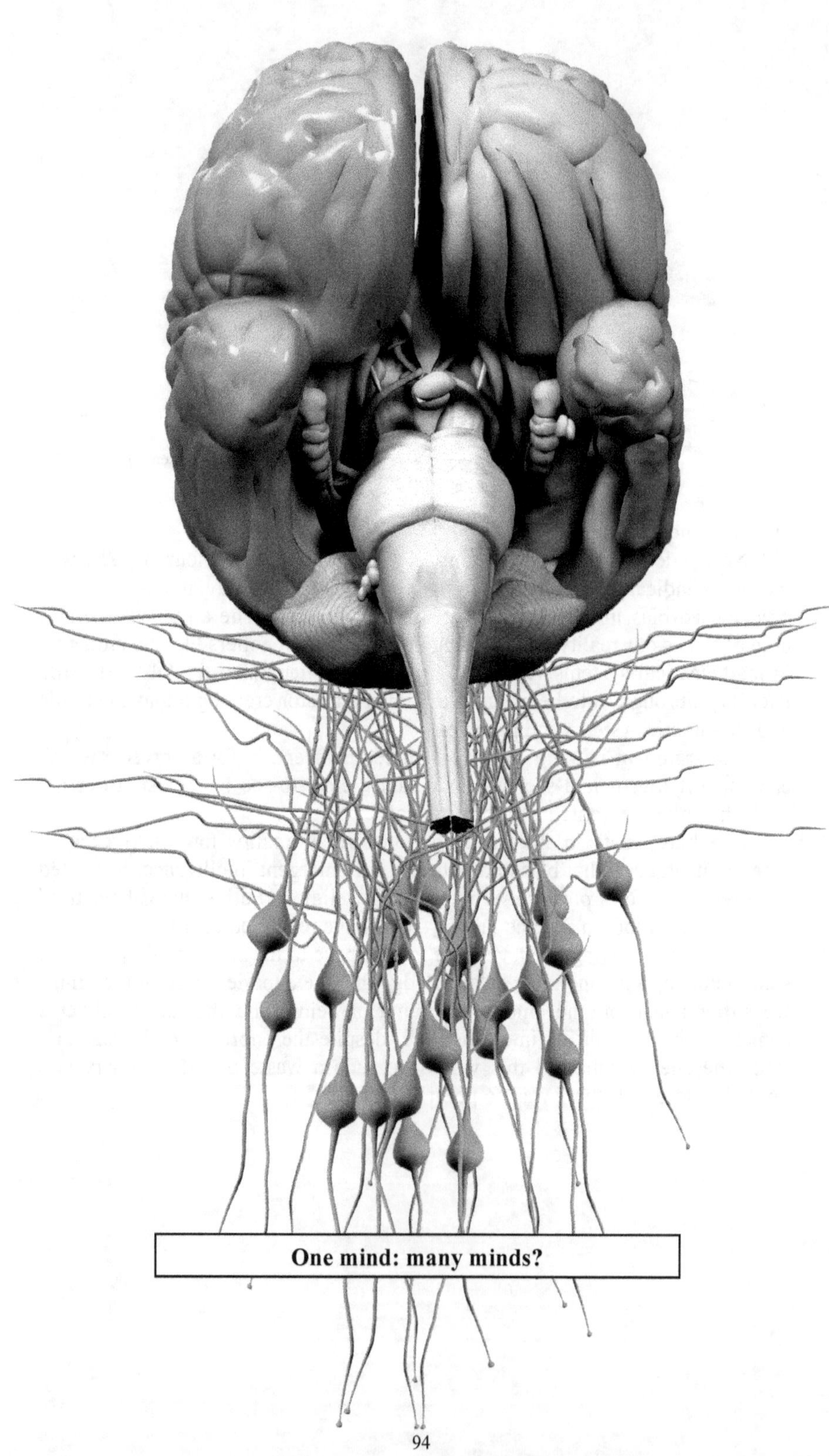

One mind: many minds?

Chapter 8: One mind: many minds?

There are some topics where events and reported phenomena cannot easily be replicated under strict scientific scrutiny. Near death (post death) experiences being one of a number of such topics. Many people seemingly die during cardiac arrest or during surgical procedures, Despite being considered dead—and there are varying degrees to what death of a person actually is—they are successfully resuscitated. Many of them report quite similar events. Some people report they recall hovering over their own bodies watching or hearing the medical staff in attendance. Others report divine like memories of being drawn towards a very bright light and experiencing feelings of peace and love.

I think we can all see that if a mind is really just the manifestation of a living brain, dependent upon oxygen to function, it (the brain) is likely to function quite differently as oxygen supplies fail. Scientists suggest that people who experience these near death and out of body sensations are really reporting a common biological trait of a dying brain. Indeed, many psychotropic drugs exist which can produce similar effects, supporting the idea that chemical changes in the functioning parts of a brain can give rise to extraordinary and often profound conscious experience.

It doesn't matter whether you believe the mind is entirely a product of the brain, and resides solely within in, or whether you believe the mind exists as a connecting wave or field of some kind, and *it* interacts with the brain and thus gives rise to the feeling of being aware. In both scenarios, any interruption to the chemical nature of the brain would be akin to messing around inside a computer or playing with the tuning coil of an old fashioned radio. Both would cause issues with what we perceive as normal awareness.

The brain remains one of the most mysterious and complex structures. It denies comprehensive research and complete understanding. Since the invention of computers, we tend to think of a brain as a more sophisticated organic computer. But there are many who argue that even when we are able to produce artificial intelligence within a computer-based neural network, that intelligence and that hardware will still remain un-self-aware. It will not 'feel' the sensation we appear to have as humans. This is not something which can be proved until such a high-end-working artificial mind can be produced.

Other areas which have proved difficult for science to research properly are ESP, Telepathy, Clairvoyance, Déjà vu, seeing ghosts, seeing UFO's, the Placebo Effect, reports in children of past lives (reincarnation), mental illness cause and treatment, just to name some common ones. Science has not proven any of these topics to be pseudo-science, yet almost everyone engaged in scientific research and teaching would steer very wide of them. They all realise the terrible and insurmountable issues involved with subjectivity of people who witness some of the associated phenomena, and the impossibility of discerning any proof (positive or negative) in areas where scientific rigour and experiment cannot be applied.

Prior to Islam and Christianity becoming two of the most accepted religions, the idea of reincarnation existed in almost all faith systems. Maybe it's wishful thinking by people who cannot accept the finality of life ending in death? A belief that you have a soul which survives biological death and possibly is born again within another new baby (or animal) gives hope and support to the dying and their loved ones. Christianity and Islam replaced the hope of coming back with one of going to a paradise called heaven, where you live forever in some kind of ideal state of happiness.

Some faiths believe in a mix of these two notions—that you come back repeatedly until your soul escapes the lure of earthly and physical life to raise itself up to a higher plane of existence.

It is quite interesting to note that the older belief of reincarnation does not seem to require the existence of a God. The newer religions do, for who else will run the kingdom of heaven?

The ideas of Christianity and Islam also seem to be male-centric. God made man in his own image... a man. The female gods of older faiths are ruled out in one hit. Faiths exist most strongly where the contemporary and rational objective vision of science is weak or non-existent. Many people today have turned away from religious worship but not necessarily from beliefs in reincarnation, or the idea of a god, or indeed a heaven. The conditioning we receive when we are very young sticks around forever.

One might also suggest the reason people do still retain a belief in 'something' and suspect life is not purposeless, is because they comprehend the way faiths in the past quickly became corrupted by people seeking to control great swathes of human population, and by the champions of a religion behaving contrary to the principles of said faith, and often waging war or worse against other groups of people: *"Ready to go over the top, lads? Remember, God is on our side!"*

We are always guilty of reframing higher thought and abstract ideas back into human mediocrity, and exploiting them within a human hierarchy. Given all the various faith systems humans have and the progress of science to probe our physical world successfully, people still remain unhappy and fearful of their mortality. Whereas once, a faith may have offered some hope when facing the abyss, the science we celebrate as improving our physical lives, removes any hope what-so-ever of surviving death. Since we are all mortal, science has not made us much happier in this respect.

The human soul

For religious faith to deliver any hope of surviving mortal death, the idea of a soul was born. A part of you, invisible, un-weighable, undetectable, resides within your body or mind. Upon death of your physical form, this soul is released to either find another lump of organic mass in which to be born again, or to move onto a paradise or an eternal hell.

It is this notion of a soul, a self without physical structure, which creates all the issues between scientific realisation of who we are and the religious or philosophical views. In all aspects of science, we only accept that which can be measured. Attempts were made during Victorian times to detect minute weight loss at the time of death; the idea being that the soul had physical properties, like mass. A dying man's bed was set up onto a contraption of weights and balances in order to detect the slight difference in weight, if any existed, at the moment of death. No discernible weight change was recorded. But the idea of a soul was conceived centuries before Maxwell and other innovative researchers discovered magnetic and electric fields in the 17th century, and prior to the foundation of quantum physics in the 21st century.

Before modern science, the idea of a soul and indeed, anything else which existed in the human form, either based on material or energy (like the chemically created electricity of neurons and nerves), did not require explanation of how something within a living person might survive his/her demise on earth.

What is required to postulate a soul idea is a new hypothesis—one which might also help to explain a mechanism for other apparent paranormal experiences many people have. The idea of self begins with a sense of self-awareness, hunger, discomfort, pain, warmth... these are the initial sensations of a new born baby. Even before we can see, we are hungry. We have no idea of what ails us, until the milk flows and removes that discomfort and satisfies a need. Our physical selves in the universe have to maintain their structures in order to have something here with which to experience. The idea of a unit of self is born in those first moments and are enhanced by all the other physical interactions we experience when encountering a physical universe.

However, we could conjecture that our true selves (or self?) is not based here at all—not in our universe and not in our bodies. We could exist in a quite different type of reality, one constructed of completely different material forms, or types of energy; 'stuff' which does not exist in our universe because they were never created in the Big Bang. And being there, we might also somehow interact with ourselves here!

What would then be required is a way for our true selves (or self?) to interact with the organic matter in this universe and only where that matter can create a neural net (a mind/brain enabling form) for our interacting field or energy to connect with. It might be impossible to detect such a field using any of our instruments to date. It might be impossible to erect any kind of other field or structure around any test subject such that our real self field can be cut off from the physical self.

Of course, we have swapped one set of unprovable mechanisms for another. Instead of an un-weighable soul, we now have an undetectable wave or field confounding any proof or disproof. But there are advantages to this hypothesis over the individual soul one. As a field, we exist before, during, and after the death of a brain here. Our minds will be a combination of what we really are in the other reality/universe and the experiences caused by our interactions with an organic brain here in this universe. It may also answer the question of reincarnation. When a brain dies here, the field may be able to interface with a new unborn mind. Seeing ghosts could be memory echoes coming into your brain from the field you are part of, and something failing to be filtered out by your biological system.

This is all ifs and maybes, I know. And in the absence of any way to test and prove any such hypotheses, it cannot be an idea acceptable to the present scientific rigours required to develop the idea seriously.

Is there anything we can detect in this reality which lends further evidence such a mechanism may exist at all?

An interesting component of our physical universe is the particle called a neutrino. It is an electrically neutral, weakly interacting, elementary subatomic particle with half-integer spin. The neutrino (meaning "little neutral one" in Italian) is denoted by the Greek letter ν (nu). All evidence suggests that neutrinos have mass but the upper bounds established for their mass are tiny even by the standards of subatomic particles. They just pass through everything, including the earth itself, your body, and all electric and magnetic fields. Originally, it was believed they were completely devoid of all mass. But more recent models suggest they have the tiniest hint of mass which means they can interact with physical matter, although it is at a level and magnitude too small to detect.

Neutrinos do not carry electric charge, which means that they are not affected by the electromagnetic forces that act on charged particles such as electrons and protons. Neutrinos are affected only by the weak sub-atomic force, of much shorter range than electromagnetism and gravity, which is relatively weak on the subatomic scale. Therefore a typical neutrino passes through normal matter unimpeded.

Neutrinos are created as a result of certain types of radioactive decay, or nuclear reactions such as those that take place in the Sun, in nuclear reactors, or when cosmic rays hit atoms. There are three types, or "flavours", of neutrinos: electron neutrinos, muon neutrinos and tau neutrinos. Each type is associated with an antiparticle, called an "antineutrino", which also has neutral electric charge and half-integer spin. Whether or not the neutrino and its corresponding antineutrino are identical particles has not yet been resolved, even though the antineutrino has an opposite chirality to the neutrino.

Most neutrinos passing through the Earth emanate from the Sun. About 65 billion (6.5×10^{10}) solar neutrinos per second pass through every square centimetre perpendicular to the direction of the Sun in the region of the Earth.

These particles are extremely difficult to detect. In fact, directly detecting them is impossible! Neutrinos cannot be detected so easily, because they do not ionize the materials they are passing through (they do not carry electric charge or many other proposed effects). They do not produce traceable radiation. A unique reaction to identify antineutrinos, sometimes referred to as inverse beta decay, as applied by Reines and Cowan, requires a very large detector in order to detect a significant number of neutrinos. All detection methods require the neutrinos to carry a minimum threshold energy. So far, there is no detection method for low-energy neutrinos, in the sense that potential neutrino interactions (for example by the MSW effect) cannot be uniquely distinguished from other causes. Neutrino detectors are often built underground in order to isolate the detector from cosmic rays and other background radiation.

So, here, quite recently, we discover a weakly interacting particle in our universe. Maybe something exists even more difficult to detect—a quantum wave that fluctuates through all universes occupying similar space, or a field which interacts between sub-atomic particles in each parallel universe (as in the set of multiverse conjectures).

If the real self, we each might actually be, is in some other kind of reality, we may not be too concerned by the local detail of our multiple lives—instead leaving that to be the responsibility of the human brain to manage; but we could be more interested in retaining emotional content, or a rough sketch of a cumulative set of physical experiences here. If this were true, then not a lot of information needs to be carried by such a field or wave mechanism. But, this latter conjecture may be wrong. Possibly all memory, all detail, needs to be carried away from the brain to the remote core self/selves?

No one is looking for such a mechanism to exist. No one was looking for neutrino particles either until they were postulated first by Wolfgang Pauli in 1930 to explain how beta decay could conserve energy, momentum, and angular momentum (spin). In contrast to Niels Bohr, who proposed a statistical version of the conservation laws to explain the event, Pauli hypothesized an undetected particle he called a "neutrino" in keeping with convention employed for naming both the proton and the electron, which in 1930 were known to be respective products for alpha and beta decay. He considered the new particle was emitted from the nucleus together with the electron or beta particle in the process of beta decay.

Without science maintaining a collective curiosity in the idea of what a mind actually is, and accepting some form of breadth in its rigours and its political aversion to investigate things which do not readily fall into an umbrella philosophy of a materialistic reality, we will never consider a possibility of mind, memory fields, or minds elsewhere might be interconnected with our 'apparent' minds here.

The more profound idea is only one field or mind may exist, not one per person. A collective mind reaches out through all possible realities, experiencing a multitude of possibilities, events, emotions, realisations, and learning. Why such a collective mind should do this, might be no more difficult to explain than by suggesting it is done just to experience novelty rather than boredom. What do you do when you are too tired to watch TV or work, but not tired enough to sleep? Daydream?

Such an hypothesis more readily explains the dying brain experience too. Those people who hearts stop, or whose brain function seems to fade to nil in serious operations, and who come back and report similar esoteric experiences, may not be witnessing their souls floating above the operating table, nor a blinding light of heaven; it might instead be the 'disconnection' experience between the singular local brain and the distant real mind. Freed from the limitations of perceiving reality due to a physical presentation here, you once again experience the whole real thing!

The oneness of being

People who die and come back again often report a sensation of mind expansion. They speak of feeling being part of everything and sensing everything is part of a whole. They talk of sensations of overwhelming love, calm, and security. People not experiencing the death event can often report momentary incidents in their lives where they become suddenly over-whelmed by a similar effect and say they have had a divine experience or revelation. Many hallucinogenic drugs can create similar sensations and states of consciousness in people too: LSD, Ecstasy, DMT.

While the exact mechanisms by which hallucinogens and dissociative drugs cause, their effects are not yet clearly understood. Research suggests they work at least partially by temporarily disrupting communication between neurotransmitter systems throughout the brain and the spinal cord regulating mood, sensory perception, sleep, hunger, body temperature, sexual behaviour, and muscle control.

One drug can lead to what users claim to be a common experience. It's frequently called 'The God Drug'—DMT!

From Wikipedia...

N,N-Dimethyltryptamine (DMT or N,N-DMT) is a psychedelic compound of the tryptamine family. Since DMT resembles the basic structure of neurotransmitters, when ingested, DMT is able to cross the human blood-brain -barrier, allowing it to act as a powerful hallucinogenic drug that dramatically affects human consciousness. Depending on the dose and method of administration, its subjective effects can range from short-lived, milder psychedelic states to powerful immersive experiences; these are often described as a total loss of connection to external reality and an experience of encountering indescribable spiritual/alien beings and realms. Indigenous Amazonian Amerindian cultures consume DMT as the primary psychoactive chemical (one that affects the mind) and is used as a shamanistic brew for divinatory and healing purposes. In terms of pharmacology, ayahuasca combines DMT with a MAOI, an enzyme inhibitor that allows DMT to be orally active. Its presence is widespread throughout the plant kingdom. DMT occurs in trace amounts in mammals, where it functions as a neurotransmitter and putatively as a neuromodulator. DMT is also produced in humans; however, its production and purpose in the brain has yet to be proven or understood. It is originally derived from the essential amino acid tryptophan and ultimately produced by the enzyme INMT during normal metabolism. The significance of its widespread natural presence remains undetermined. DMT is structurally analogous to the neurotransmitter serotonin (5-HT) and the hormone melatonin, and furthermore functionally analogous to other psychedelic tryptamines, such as 5-MeO-DMT, bufotenin, psilocin, and psilocybin.

Note that all plants and most mammals contain this molecule, including us. I find it quite strange that the introduction of a tiny amount of a fairly simple molecular compound can have such a powerful influence on our conscious grasp of reality. The normal effect is journeying to another place, one akin to creativity and exotic possibilities. People who have taken the drug report it is **not** like being drunk or out of control. They say "All your baggage comes with you". They insist they have all their facilities intact and it feels like taking a journey through a tunnel or worm hole to a different place.

Maybe, when we speak of a drug like DMT shifting consciousness, we should wonder how the current set of naturally produced drugs inside our brains are limiting, or filtering out other possible states of consciousness. To me, a drug like DMT seems to break down a naturally occurring set of chemical filters designed to focus a mind in a brain only in the perception of this reality, but once the inhibitors are removed temporarily, the mind is free of such constraints.

Unfortunately, these drugs, originally being used as part of scientific research into the question of consciousness and to help understand mental disorders, started to be used increasingly for social recreation during the 1960s and onwards, predominantly by teenagers. Theie use spread until DMT was considered a threat to political control systems in western countries. If people started feeling a 'oneness' with other people, plants, the universe itself, they are left with a legacy of that experience and insight. People who become enlightened by the idea that people all over the planet are not so different to themselves are far less willing to put on a uniform, pick up a gun, and start firing bullets to kill other people through an act of war.

The drugs were made illegal. Medical and chemical research ceased and became frowned upon. It's a damn shame, because controlled use of some of these drugs are likely to lead to a more comprehensive understanding of what consciousness and mind is. Beyond the major powers of western countries, the use of many of the hallucinogenic and disassociating drugs is still allowed. But you will need to travel to South America to try out the DMT experience yourself.

I think my point in all this is, the human mind can experience a wealth of different states of awareness besides the everyday one we probably commonly share. It's also interesting to ponder why every plant has DMT in its make up. DMT is a neurotransmitter-affecting chemical. Plants don't have neurotransmitters. They don't have brains. The exotic question is, are they experiencing something akin to a kind of awareness outside our realm of understanding. Are all living things in some way in contact with a bigger picture of reality. I wonder sometimes: could they be more 'connected' than we are?

The ultimate experience

I hesitate to include this section. First of all it's from a set of personal experiences, and I'm concerned relating them might weaken my work here and you, the reader, might categorise my good self as a madman. It is so easy to miss the detail in the explanations and instead make generalised and incorrect observations. However, I'm going to take a gamble on your astuteness, and your capability to differentiate between a raving loony and a sober man whose thoughts and ideas have been partly shaped by the curiosity springing from some of his own experiences.

First of all, I have, and do suffer a form of mental illness. (*Don't shut the book yet!*). I hadn't realised I had this specific weakness until I was 26 years old. I am 64 now. It did present itself at an earlier time, which I will describe in a moment, but I hadn't realised it back then. In fact, it took me half my life to understand the cause and effect of what often comes to plague me.

To set the tone and to hopefully give weight and credibility to a particular experience I once had, and one I wish to share with you, I must first qualify my particular aspect of mental flaw. I think the best way is to tell you how this flaw gradually dawned on me.

It's first manifestation was when I was sixteen years old. I was in the period of taking GCE's, a set of qualifications at secondary school in England. These exams are important. Success means entry into higher levels of (back then—*free*) education, leading to entry at University and possibilities of obtaining a degree. Failure meant you left school and took your chances along a different route. Prior to taking the exams, we took 'mocks'—similar tests at equal levels of difficulty to see where we were weak and needed to do more study. The mocks were not a qualifying set of exams. They just mimicked them. I mention here, I passed all my mock exams with flying colours.

Anyway, we are now at the time of taking the real exams. I had taken the easy one, the art exam, with the rest waiting for me after this specific weekend—a weekend, which unknown to me that Friday, was to change my entire life.

I must explain: although I was quite academic, I lived in a poor and rough area of London. School to me was almost unbearable most of the time. More activity was involved avoiding bullying and intimidation than carrying out scholarly study. Most of what I learned came from teaching myself at home. It took several years at that school before I had donned a bluffing persona which made it look to other boys that I walked the walk and talked the talk, so most of them left me alone. But a bluff it was. I was not a good fighter, and any thought of violence scared the pants off me. I just wasn't prepared to show it.

Saturday morning. Myself and my younger brother were walking along the pavement (sidewalk) just a few yards from my home. Coming towards us were two boys we were aware of but not particularly fond of. They belonged to a group of boys from another nearby neighbourhood. Like many social groups growing up, the neighbourhood seems to divide groups in poor areas. The boys were from a group who didn't like us. Needless to say, a confrontation arose, resulting in an appointment with fate. It was arranged that we meet in 30 minutes to settle our confrontation with a fight. I didn't want to fight but backing out was not an option. Losing face would echo around my school and lead to all kinds of challenges from other boys. Another and bigger issue though is I knew this boy would beat me in a fight. I had seen him play fighting with his mates many times, building up his strength and skill. I never really fought. I just bluffed. So, by agreeing, I knew I was going to be in for a lot of pain.

The alley where the fight was going to happen was near my home. I went straight home, not least to use the toilet as I was overwhelmed with fear, anxiety and apprehension. Everything in me collapsed into that fear. I searched frantically for a way out, but there was none. Any options not to fight would only lead to worse problems. The fight took place as arranged, and as I'd predicted, after twenty minutes, I was beaten and broken. My injuries were severe as I had received a terrible kick to the face. I had to go to hospital to be stitched up.

When I turned up at school Monday morning to spend the week or more taking the rest of my exams, one side of my face was so swollen, one eye was unusable. But that wasn't the problem. I seemed to struggle to get my brain into gear. Things I thought I knew seemed remote and not recallable. Each exam paper seemed to ask questions which I should know the answers to, but I had trouble remembering information and threading it together in a coherent way. The whole week was the same. I didn't even complete most of the exam papers. Needless to say, I failed all my exams except for the Art one, which I had taken before the weekend.

My father was not happy, and I was obliged to leave school and seek a job.

I always thought the exam week, and the resultant slowness of my mind, was a kind of shock reaction to the fight. Coupled with the pain I still endured from the swollen face and bruising, it seemed to be natural I was unable to perform in an optimum way. It was nine years before I learned my assumption was wrong.

I was married with a baby girl. One night I made an awful, impulsive, and regretful mistake. I betrayed my wife with another woman. God knows how it happened. A moment of weakness, desire and lust, along with too much alcohol, overcoming loyalty and love. One can blame it on whatever excuse they like. Me? I blamed myself. At heart, I am an honest and faithful man. I began to worry about the impact on my wife if she ever found out. I worried about it, thought about it day in and day out. No matter what I tried, I could not get this out of my brain. It went on for days, then weeks, then months.

I slowly discovered something was changing about me through these months. The world seemed to grow busier, faster. People started looking meaner, more selfish, hostile. I had difficulty doing my work, forgetting things I had long known. My ability at problem solving (part of my job) diminished and all my emotions save the negative ones disappeared. Physical pain began in my muscles, in my shoulders, my neck, and a tightness like a mask drew itself across the skin of my face. I could not taste anything, smell anything. Worse, slowly—my sense of touch and feeling became distant and remote. How much strength to use to open a door was no longer a natural thing, nor walking down the street. It was as if I needed to consciously speak a silent instruction to myself in my mind to carry out any kind of activity. It was as though the natural flow of thought had evaporated, leaving me with a kind of basic level thinking. The simplest of tasks became a Herculean challenge, almost impossible to achieve readily.

Visits to the doctor ultimately resulted with me volunteering to go into a mental hospital for treatment. I was there 6 months. No improvement. It was as if I was stuck in this alternative, less alive, less aware, basic state of being. My boss was about to give up on me after 6 months sick leave. I left the hospital and went back to work. I remember thinking, so long as I get there each day, without being late, I might be able to last months before they realise my work was causing more harm than good. I don't know how I survived. I went to work, did a poor job of it, came home, went to bed. Got up, went to work, came home, went to bed. Day after day, week after week—month after month.

It took 18 months before that altered state lifted. It was the worse 18 months I have ever lived. Nothing in my life from that moment to this has ever been so grim and painful.

The doctors' verdict? Depression, brought about as a result of anxiety! I have had 26 or more similar such incidents since. Some last weeks, some last months. A tiny, almost insignificant worry which I get locked into can bring it on. Sometimes, if I catch the signs early, I can take avoiding action. Once depressed, no pill clears it, but if I catch the anxiety at the right time before it maximises and I take one of the pills for a few days, I can sometimes prevent the depression altogether or minimise the length of time it lasts.

My view on it is that we all have to go through anxiety, stress, and fear. It's part of life. If I can remove the worry, solve what's causing my anxiety, I don't get ill. But there are some things you just can't solve quickly or you have no satisfactory solution for.

I don't get periods of high activity. I do not suffer from manic depression, also called a bi-polar disorder. I have revealed to you this weakness so you can understand I am a person who acknowledges how a mind, my mind, can experience different states of awareness. For what I am about to say next needs to be put into its correct context. I had this experience which I never had before and I have never had since. During my life, I have experienced so many things, but this... this was by far the most profound.

I was 50 years old. I had gone through a bad period of my life for about two years, resulting in me leaving my safe job, my wife, and my home. It wasn't something done easily and there are lots of reasons why these events happened without placing blame anywhere. I would have preferred not to have done it if the issues driving me into repeated depressive bouts could have been fixed. They couldn't, so my solution was to leave all that was contributing to my pain and suffering. I didn't have a plan on how to build another life. Sometimes, things just get so tough, you just have to take action and see how it goes from there on in.

I ended up living in a room in a warehouse. No windows, no bathroom, no kitchen. It was difficult for about 6 months. I went into a depression over the worry of it all and came out of it again. Then it wasn't so bad. I formulated a plan and started a small business. I had savings and while they lasted, I could keep the show on the road until my business brought in sufficient funds. In many ways, I felt satisfied to be free of the things which were all making my life hell. This wasn't heaven, but at least I could think straight again consistently to work my way out of the situation. I still had worries, but not at the level which causes me anxiety. I could see solutions to all the things that troubled me. Thus, the anxiety was at a level not high enough to trigger another bout of depression.

My brother went on holiday with his wife and had offered me the opportunity to stay at his house and mind it while they were away. A bit of comfort would be good, so I had accepted the offer. I slept there 2 nights, returning to the warehouse room in the day to carry on with my work. The third night I went into the house and took a shower. I had one beer and went out into his garden. It was one of those warm, sultry, balmy English summer evenings—clear dark sky filled with stars, warm blossom-scented air. A rare English evening.

I lay on the lawn and stared up at the stars above. I was looking for maybe a minute or so, and then it happened. How I explain it is weak compared to what my real experience was like, but words cannot express the moment well.

The stars fell into me. One minute, they were there glittering and shimmering like bright jewels set in a blue-black immensity. Next second, they were rushing towards me, funnelling into a rapid stream that whacked into my chest. I had a moment of apprehension. I could see this super white torrent of light was charged with energy, and it was about to strike me. And then it did.

Initially, I thought I was being electrocuted—that they hadn't been stars after all, but a lightning bolt. Then I realised I wasn't. It was more a rush of excitement like an intense high, like orgasm, or what I assume addicts get when they inject opiates. With the rush, came something. I'd heard said, but never understood the phrase. My mind expanded!

For a moment, I felt as if my mind reached out and touched the very boundaries of space and time, to the far edges of the universe. I knew I was in a single moment comprehending the reason for everything... like I had pulled my head out from a fog, and could see again. There was a notion (not a voice) that the entire universe and I were one, and now we were one, I could sense its thoughts as though they were my thoughts, and if I could put those thoughts into words, they were saying, "Everything is safe. Everything is as it should be. Nothing dies."

All the time, I felt this overwhelming sense of safety, love, and warmth, as if these were the only sensations that existed. I didn't meet God, Angels, go to heaven. None of those things. I experienced love in its absolute form, love between all components of the universe, a nominal sentience formed by all that existed in it... or something like that.

And then it ended.

And I was lying on the grass looking up at the stars wondering what had just happened. At first I thought I might have had a mini-stroke, then I thought maybe I had fallen asleep for a few seconds. Both of these could be true and make more sense than any supernatural explanation. I know, like you know, stars do not fall from the sky and pass through human bodies. The physical world out there had done nothing at all, but my mind had manifested this experience. The cause may well be unimportant. Stroke, nap, or some kind of tiny malfunction in the way parts of the mind are filtered from other parts—none of this mattered.

What counted was that I had experienced what it's like to have a mind unfettered by constraints, a mind which might be able to exist without dependence on biological control systems and organic hardware, a mind that doesn't need to build an internal mirror of the outside reality, but instead be part of and integral to it. I had experienced the concept of universal mind.

In later years, I read of many other people having similar experiences, some on drugs, some not. I have read quite reasonable scientific explanations of a cause for this experience when someone takes mind altering drugs. I'm not really interested in the cause for my own unique experience. I am interested in the experience itself. In the moment I felt like I knew the answer to everything, anything. There was this sensation of it all being so easy and obvious to understand. There was an abstract thought like a multi-thought which seemed to deliver answers to all questions simultaneously, like time had become meaningless and that all thoughts could be experienced at the same time, instead of consecutively. Even this, now I read my words, is a poor description, but try and explain the taste of an apple or the fragrance of a spice, especially to someone who has no taste or sense of smell. Words cannot explain things which cannot reference similar content experienced by someone else.

It would be easy for me having been raised in a predominately Christian country to suggest that what I had felt was the mind of God. But it was not like that. It was more like the mind of the universe was entangled with my mind, and your mind—every living thing was a sub-part of that mind.

None of this is any way scientific. There is no witness to say if I fell asleep or not, no witness to determine if I underwent a mini-stroke. I am not asking anyone to explore the cause. I am asking everyone to accept such a sensation without drugs can be experienced, and once experienced—it will leave a powerful set of notions on the person who has it. The most significant thing is, no other state of awareness can come close to the utter and complete safety and wonder I felt in that brief moment. If we knew this is what it feels like to die and let the mind free of its filtered entanglement with our human brains, we would all be leaping under cars and jumping from buildings to get there straight away.

We can debate forever whether or not that experience is any way a glimpse of something truer than what we can normally perceive. We can argue it is, or is not, a product of brain chemistry going awry, or me falling asleep. But then, isn't my entire understanding of the universe and myself no more than the waking dream of my brain and its chemistry anyway?

* * * *

Where do we live now?

Chapter 9: Where do we live now?

To understand your own bit of reality, you know—the mirror version of the real one outside your head—to store an accurate picture of not just your earthly world alone; to really start glimpsing the absolute absurdity of materialistic-only thinking, you need to look out further into the cosmos, and place your physical self within the enormity of all there seems to be.

The image below is the most resolved photograph of our galaxy ever taken—the Milky Way, as photographed from earth and looking edgewise though the spiral cluster towards the centre. What you see here is the core of

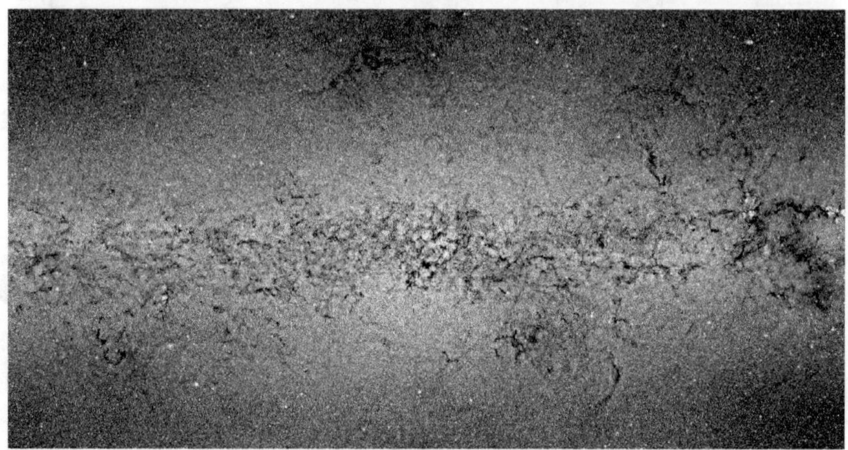

the our galaxy, as seen by the European Space Agency's VISTA telescope. If you looked up at the centre of the Milky Way with your naked eye, these stars and dust clouds would occupy a patch of space that's just a few square inches. This is the most detailed photo ever captured of the Milky Way, enabling ESO astronomers to catalogue no less than 84 million stars.

The source image has a resolution of 108,500×81,500 pixels or 9 Giga pixels, and is 24.6 gigabytes in size. To reach such an utterly crazy resolution, the VISTA telescope took thousands of photos of the sky, and then compiled them into this single, 9-gigapixel mosaic. VISTA, in case you were wondering, stands for Visible and Infrared Survey Telescope for Astronomy, and with a 4.1-meter mirror—it is the largest visible and near-infrared survey telescope in the world. Put simply, this is the best view of the Milky Way ever. Of course, it's a tiny little thumbnail here but you can see it in all of its stunning resolution at: *http://www.eso.org/public/images/eso1242a/zoomable/*

It was once thought there might be up to 17 billion possible habitable planets in our galaxy but recently, astronomers at the University of Auckland claim that there are actually around 100 billion habitable, Earth-like planets in the Milky Way—significantly more than the previous estimate. Current estimates suggest there are roughly 500 billion galaxies in the universe, meaning there is somewhere in the region of 50,000,000,000,000,000,000,000 (5×10^{22}) habitable planets.

The previous figure of 17 billion Earth-like planets in the Milky Way came from the Harvard-Smithsonian Centre for Astrophysics in January, which analyzed data from the Kepler space observatory. Kepler essentially measures the dimming (apparent magnitude) of stars as planets transit in front of them—the more a star dims, the larger the planet. The University of Auckland's technique, called gravitational micro-lensing, measures the number of Earth-size planets that orbit at twice the Sun-Earth distance. This results in a list of planets that are generally cooler than Earth—but by interpolating between this new list, and Kepler's list, the New Zealand astronomers hope to generate a more accurate list of habitable, Earth-like planets, and anticipate a number in the order of 100 billion.

Suffice to say, if the Milky Way contains 100 billion Earth-like planets, and there's somewhere in the region of 500 billion galaxies, then there's an extremely high chance of other planets harbouring life. How we'll get to those planets, or alternatively—how the residents of those planets will get to us—remains a very big question. The nearest probable habitable planet is Tau Ceti e, which is 11.9 light years from Earth. The fastest spacecraft ever, Helios II, travelled at 43 miles per second (70km/s), or 0.000234c (the speed of light). At that speed it would take 51,000 years for a spacecraft to reach Tau Ceti e.

It's highly likely many possible intelligent forms of life exist throughout the universe based on the fact of life being here on this one. Maybe, better said, it's highly likely intelligent life emerges on many planets on many systems throughout the universe, but they might not last as technologically capable beings for very long. As technological advances are made, unless those advances are made alongside an equally evolving cultural system—one which binds intelligent individuals socially and eradicates war—then the chances are war itself will reduce those civilisations back to the stone age.

Let's forget about any possible other life forms for now and just focus on you living and positioned on earth, in the Milk Way galaxy. Galaxies come in various shapes and forms. We cannot actually photograph ours from a "bird's eye" view, because to do that, we need to be outside of it. The next best thing is to pick another spiral galaxy roughly the size of our one and use that as a representation of the Milky way.

I have printed an image overleaf of such a galaxy, courtesy of Nasa and their public domain images. It is NGC 3949, photographed by the Hubble telescope. Like our Milky Way, this galaxy has a blue disk of young stars peppered with bright pink star-birth regions. In contrast to the blue disk, the bright central bulge is made up of mostly older, redder stars. NGC 3949 lies about 50 million light-years from Earth. It is a member of a loose cluster of some six or seven dozens of galaxies located in the direction of the Big Dipper, in the constellation Ursa Major (the Great Bear).

If this were our Galaxy, I would make a rough estimate of where our solar system (our sun and planets) would be in one of the arms (such as I did in the image overleaf!). No one can say for certain, but at the very bright centre of the galaxy, there is probably a black hole, a super dense gravitational centre which holds all the clusters in gravitational rotation.

You really have to consider this and not let it escape your command of position, size, and scale. Many people, understandably, will be more ...

Milky Way type galaxy. Image Credit: NASA/ESA/Hubble Heritage Team

...concerned, and have more thoughts about the immediacy of their earthly lives. Same here! But they, you, me have a silent moment here and there. We have an opportunity to think outside of our 'little' fights, and our "little" needs, and the other things we do to maintain our grasp on life.

Within that spinning whirl of light, I suspect, just like us, there are other living sentient life forms who might look out at their skies and ponder the same. You will never meet them, not now, not at this time. The only way you might reach them is through the wonderment of, and by the pondering of, these ideas. Your bond with them will be though an abstract: the shared wonderment. You may for a brief moment connect with them through the sheer and emotional aspect of the empathy within you and the same, if it exists, in them. You might even get a shiver up your spine!

Once we looked out, 2000 years ago, and saw a multitude of gods there, and us the pawns of their game-play. Today we can look out and wonder if intelligent forms might be fighting their own demons of being alive and lost in stardust, and just getting by—like us.

A galaxy and us

Size matters. You see an ant in your garden, or a tiny spider on the floor. They are so small and you are so big. You are made of cells. Your entire construction as an adult human being is a self-organising, self-co-operating mass of some $1x10^{14}$ cells of various size and type. To position this in a comprehensible framework, it amounts to some 100 trillion cells, which itself is 1000,000,000,000 cells, give a nought or so! Each cell of your mass is made of atoms, just like the atoms of those star systems in that galaxy above. Each cell in your body contains (by an odd coincidence) 100 trillion atoms. This is represented by 10^{14} atoms in a cell. Your entire mass is 10^{14} x 10^{14} atoms all working to a collaborative set of rules and order (bottom up) to keep you alive. The atoms have been collected over a period of some 13.5 to 14 billion years to coalesce you into position in a vast and bewildering entity we call reality. So, the universe fitted together 10^{28} atoms, obliging them to cooperate, for you to be here right now!

Why should it bother?

Purposeless?

You really have to suck this in. It is extraordinary. Many paradigms of thought and various descriptions about this unthinkable truth miss the mark and reduce it to an aspect of universal and blind natural processes. Yes. I agree it's natural. Why? Why would it be natural for a vast universe to collect its resources together from an original and brief burst of energy— an event so powerful it brings everything into being—and then redefine parts of itself as swirling masses of galaxies, and despite the 'seeming' chaos and confusion, place you delicately in the arm of a spinning gigantic whirl of energy and matter, so dangerous, that if it even winked at you, it would blow you out of existence?

Do you not think, even for a second, in a detached and possibly emotional way, you are not extremely privileged? Lots of galaxies. Lots of stars. Lots of planets. Lots of living beings?

All this power. All this energy. All this vastness. Have a look again at the picture on the left. Look at the end of that little pointed arrow. In the picture on the previous page. This is where you are.

It does not matter if God or the devil put you there. It is equally unimportant if a blind and non-sentient universe put you there. That's your position right now. Had the dinosaurs been able to consider it and move on, they might still be here and not us. But they couldn't. They are all dead. They are dust. You are made from the recycling of their dust, their embers, their inability to comprehend their position within the bigger picture. You are the next gamble of a universe which has placed you here briefly, safely, by whatever means and process, to consider the puzzle and the bigger picture.

You are probably part of Plan B! And you, along with any other intelligent life—if it exists—elsewhere, is all there is between the unknown consequence of whatever is going on in reality (this universe), and its end.

Your intelligence, your outward mind and curiosity, despite the instincts and natural processes you inherited to reproduce, stay alive, fight off the confusing completion in a smaller environment—they are the only difference between matter being just that, and matter possessing awareness.

Something is going on. Something outside of our easy to understand local existence. It seems to me a good idea to recognise this. Not least, if we just get by with the local stuff—eat, sleep, reproduce, we will just end up as a failed Plan B and go the way of the dinosaurs. I hope then, in that case, there is a plan C.

Machines?

* * * *

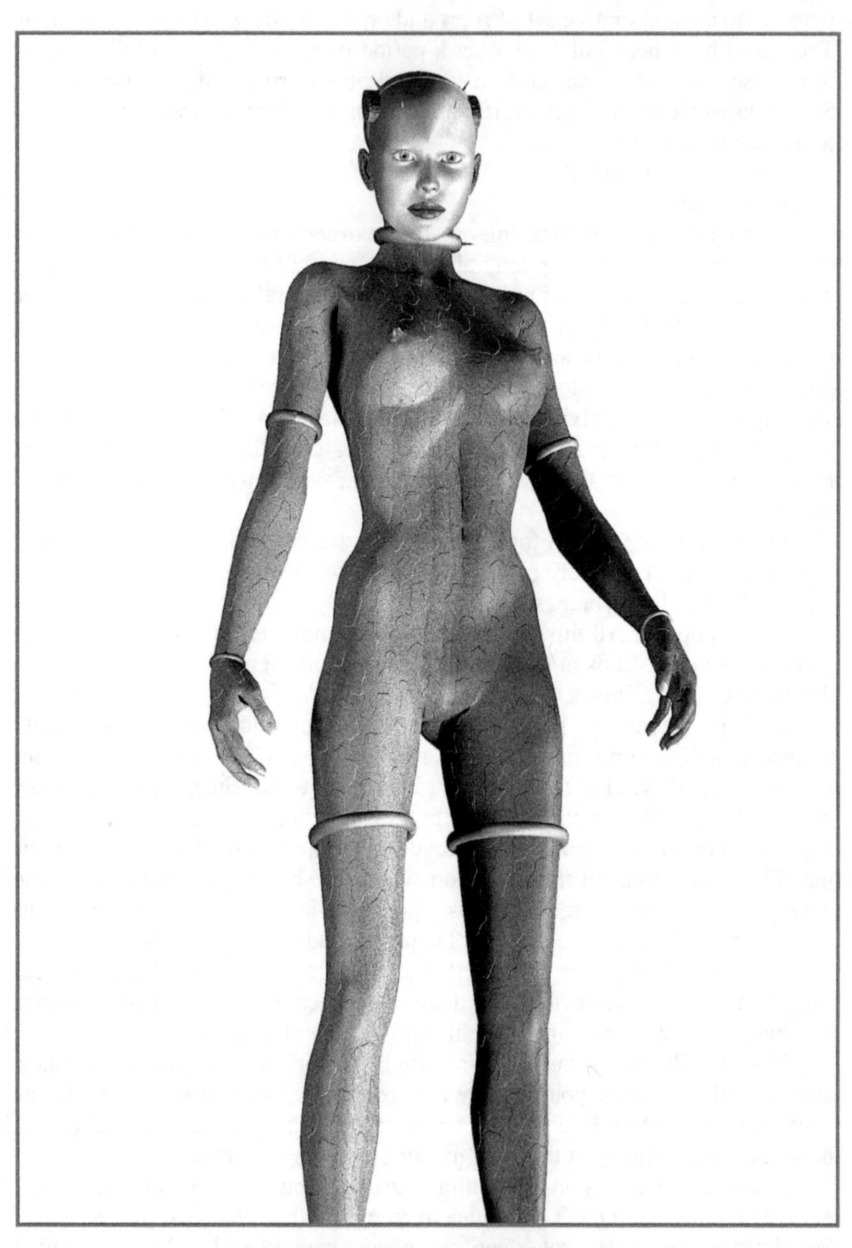

Rise of the machines?

Chapter 10: Rise of the machines?

The advance in computer technologies has been staggering. In just thirty years, computers have replaced a multitude of people's work roles in society and transformed our lives in an unprecedented way. And there is no going back. Today, we shop on line, bank on line, meet new partners through dating web sites, read books and newspapers on electronic devices. Our phones have become mini computers, able to track our location on a permanent basis. Social networks trawl our messages to friends and our postings to submit data to software which determines our likes and dislikes in order to push advertising for products and services at us.

Our cars are managed by computers. The planes we trust as being piloted by a human being to fly us off to exotic holiday destinations are un-flyable without a computer interpreting and correcting every move the pilot makes. Our very lives are now dependent upon silicon chips, copper wires, and sophisticated software programs.

Computers are performing faster, smarter, and their capability is advancing logarithmically. Many people wonder how long it will be before they replace us!

Is this a real possibility? On the one side, we have a set of professionals who argue computers will always be under our control. On the other side, there are others who argue it's just a matter of time before computers, and neural networks running on them, bring about self organising aware machines. My own thoughts tend to lie with the latter arguments. Machines will become aware. More—they will become self aware.

I think it will take longer than predicted, but certainly all the activity and use of computers to intersect with our lives, and our increased dependence on them, will demand a slow evolution of their finer capabilities. It's just a question of time before they, the intelligent machines, gain enough processing power to start to form a kind of consciousness, probably superior to our own. At the time of writing, the world's most powerful super computer is China's Tianhe-2, also known as the Milky Way-2. In June 2014, Tianhe-2 remains on top in a list of supercomputers at the speed of 33.86 Peta-flops in HPL Linpack benchmark. A petaflop is the ability of a computer to do one quadrillion floating point operations per second (FLOPS). Additionally, a petaflop can be measured as one thousand teraflops.

On the second fastest computer, scientists ran a simulation program emulating 1.73 billion virtual nerve cells and 10.4 trillion synapses, each of which contained 24 bytes of memory. The simulation took 40 minutes of real, "biological" time to produce one virtual second.

Billions and trillions of simulated neurons and synapses are nothing to sneeze at, but keep in mind how that equates to only one percent of what's going on in our brains. Our biological super-computer and neural net, our brain, by comparison, consists of about 86 billion neurons linked together by trillions of synapses, making for a total of hundreds of trillions of different pathways for brain signals to travel through. That's a lot of electrical impulses shooting through the brain at once, which means a lot of processing power.

It seems computers still have a way to go to begin to match what nature has created in us. If the advances continue, what power will computers have in say just 100 years from now? Quantum computers are already in their infancy and offer a potential leap in the order of several magnitudes to current computing power.

Organic Computing has emerged recently as a challenging vision for future information processing systems. Organic Computing is based on the insight that we will soon be surrounded by large collections of autonomous systems, which are equipped with sensors and actuators, aware of their environment, communicate freely, and organize themselves in order to perform the actions and required.

The presence of networks of intelligent systems in our environment opens fascinating application areas but, at the same time, bears the problem of their controllability. Hence, we have to construct such systems—which we increasingly depend on—to be robust, safe, flexible, and trustworthy as possible. In particular, a strong orientation towards human needs as opposed to a pure implementation of the technologically possible seems absolutely central. In order to achieve these goals, our technical systems will have to act more independently, flexibly, and autonomously, i.e. they will have to exhibit life-like properties. We call those systems "organic". Hence, an "Organic Computing System" is a technical system; one which adapts dynamically to the current conditions of its environment. It will be self-organizing, self-configuring, self-optimizing, self-healing, self-protecting, self-explaining, and context-aware.

The vision of Organic Computing and its fundamental concepts arose independently in different research areas like Neuroscience, Molecular Biology, and Computer Engineering.

Self-organizing systems have been studied for quite some time by mathematicians, sociologists, physicists, economists, and computer scientists, but so far they are almost exclusively based on strongly simplified artificial models. Central aspects of Organic Computing systems have been inspired by an analysis of information processing in biological systems.

First steps towards adaptive and self-organizing computer systems are already being undertaken.

Adaptivity, reconfigurability, emergence of new properties, and self-organization are topics in a variety of research projects. The priority research program of the German Research Foundation (DFG) is already solving fundamental challenges in the design of Organic Computing.

Computer systems of the future, especially where life critically depends upon their faultless working, will need to be equipped with self sensing and self repairing (healing) characteristics. The question is, at what point does an intelligent system become something we can say has a mind and is aware? The answer to this question might also lead to more proof of whether or not a mind is just the actions of a brain and resides only within it as the total sum of neural electro-chemical activity, or whether it is that plus its integration with a mind or species field external to the biological mass.

In all instances on earth concerning human beings, whenever a more intelligent group has encountered a less intelligent one, mostly in terms of that intelligence having access to more powerful technologies, the brighter bunch has wiped out the lesser one, or absorbed what's left of it into the brighter group's culture and systems. The Australian Aborigine, the New Zealand Maori, the tribes of North America, the Incas and the Aztecs are all examples where indigenous groups of people have been superseded by the more technologically advanced races who discovered their lands.

Concern about the rise of the machines might stem from a notion that the new thinking machines will become all too aware of our human flaws. We fight wars. We serve basic and often primitive selfish aims. Our desire as a planet of 7 billion people is more focused on individual self-fulfilment than any kind of global and collective aim to explore the universe or achieve spiritual enlightenment. Brighter machines may suspect a better goal is achievable or simply see us humans as too flawed to take on such universal ambitions. It might be that evolution itself, having made thinking animals—us, from a starting point of single cell life forms, will somehow welcome and favour the rapid advances towards a universal goal which evades us, but one which bright new intelligences can achieve. Maybe the rise of the machines is an evitable step in evolution. Maybe the fact we were not evolved simply to pleasure ourselves, but instead were meant to serve a purpose and a quest of the universe itself—and have failed to realise it—culminates with us becoming a discarded idea?

What then of mind fields and the notion that we are not just minds in our brains but minds which *interface* with our brains? Will our external mind fields simply interface with the new non-biological brain; will our hypothetical distant whole minds elsewhere simply disconnect from a diminishing humanity, and instead reconnect with thinking self-aware machines?

Although much of this is speculative, you only have to ponder this question: what is the future for computers? The only possible answers are: they will always be machines under our control, or we cooperate with them, or they will be machines which replace us. Out of three possible choices, one is our demise. We have a one in three chance of being replaced by the thinking machines! Not good odds for survival then.

Computers are already so advanced that they are required to actually make the next generations of themselves. Without the power of one computer generation, we cannot create the next. The software managing their functionality grows exponentially more complex and sophisticated. I fear the march to a real new world, populated by smarter creations than us, has unquestionably started and is now in full stride.

In my first book, 42—The meaning of Life, The Universe, and Everything, I wrote a fictional account of how this may come about and what the future development would lead to. I'll refrain from repeating it here, but the idea is the machines go on themselves creating better kinds of awareness which no longer require a physical component system to map neural activity on. Ultimately, mind becomes an energy matrix—thinking, aware, expansive, and on such a huge scale that it can direct its own energy to physically interact with components of the universe directly; shifting universal constants, transforming matter, tapping into energy, changing gravity, interrupting whatever is causing universal expansion.

I have assumed future intellects will be free of human egos, conflicting idealisms, vanity, greed, and all the other traits which defeat any real cooperation between different human groups. My assumption may be wrong. The machines may inherit similar traits too, or possibly develop them as their self-awareness matures and becomes more refined?

Could such intelligent machines already exist out there in the Cosmos? Certainly, but we'd be very unlikely to hear from them unless our planet has resources they need and they become aware of that.

For anyone who finds the ideas of intelligent machines and the notion of them becoming self-aware, a fairytale speculation, I might say the following. According to the theory of evolution, we sapiens arose from less intelligent former biological creatures, often from forms we claim to have no real mental activity. We are machines ourselves but made from organic components. One of the theories for the development of our self-awareness is due to the complexity of managing a biological system in mammals; something which necessitated the development of a central control system—the brain. It may have started out as an unconscious system, much like the brain stem we possess today, but at some point, as the number of neurons and connections increased in a limited space, and reached a threshold—self-awareness spawned.

What nature did to us, we seem to be doing again on new materials. And there is nothing to suggest we continue only to use silicon chips. Humankind is quite capable of developing thinking networks or computer systems on organic material instead. The future is not a question of clunking lumps of titanium, plastic, and steel machines with intelligent systems managing them: it's more about a mix of organic and non-organic sophisticated creations able to self-organise and outperform us, their creators, on every level.

They may ask the same questions: "Why am I here? What is my purpose? What happens when I die?"

What applies to them will feel no different to what applies to us, except with their greater intelligence, they may be able to answer those questions and get on with a task we were evolved for, but failed to realise.

Of course, there may be no purpose. It might be that the reason for a physical reality is just novelty. There is nothing as boring as a constant state. Variability and change, multiple-experiences instead of a single one, these are ideas which appeal to all thinking people. The universe maybe no more than the product of a similar abstraction in a grander system—one also sharing our common desire for novelty.

The best humanoid robot to-date

http://asimo.honda.com/downloads/pdf/asimo-technical-information.pdf

The most advanced Humanoid type robot is a magnesium alloy robot covered with plastic resin. It was developed by Honda, the Japanese company known for cars and motorbikes. It has taken over 25 years, starting from scratch in 1987, to achieve the level of sophistication they have engineered into this extraordinary machine.

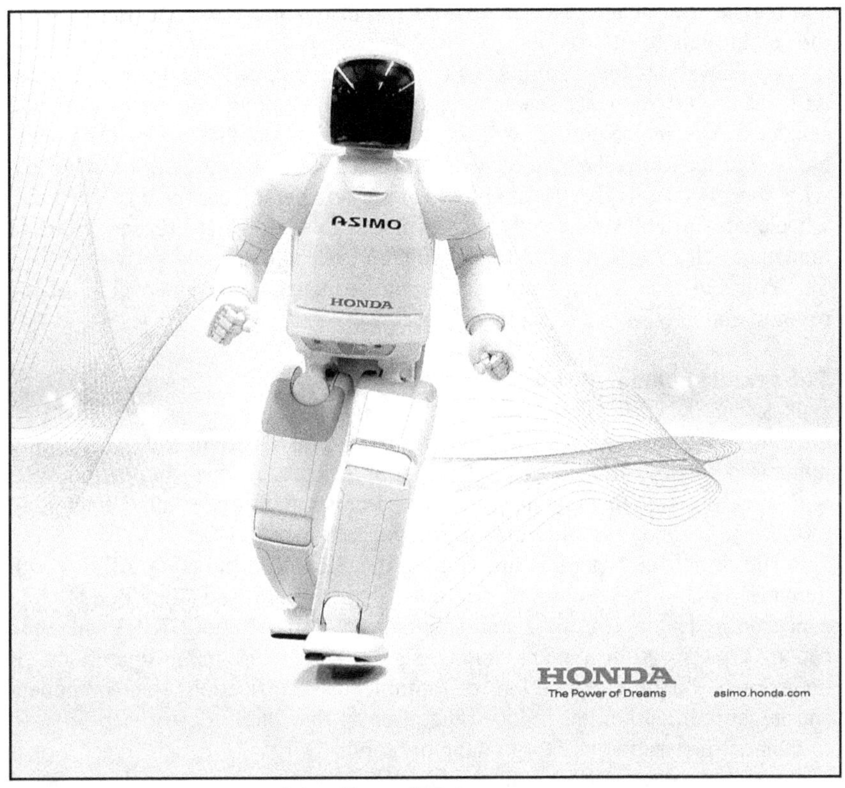

Asimo Humanoid Robot
(c) Honda. Used with permission.

The dream sounded simple. Design a robot that can duplicate the complexities of human motion and genuinely help people. An easy task? Not at all. ASIMO took more than two decades of persistent study, research, and trial and error before Honda engineers achieved their dream of creating an advanced humanoid robot.

In 1986, Honda engineers set out to create a walking robot. Early models (E1, E2, E3) focused on developing legs that could simulate the walk of a human. The next series of models (E4, E5, E6) were focused on walk stabilization and stair climbing. Next, a head, body and arms were added to the robot to improve balance and add functionality. Honda's first humanoid robot, P1 was rather rugged at 6' 2" tall, and 386 lbs. P2 improved with a more friendly design, improved walking, stair climbing/descending, and wireless automatic movements. The P3 model was even more compact, standing 5' 2" tall and weighing 287 lbs. The current version of ASIMO is the culmination of two decades of humanoid robotics research by Honda engineers. ASIMO can run, walk on uneven slopes and surfaces, turn smoothly, climb stairs, and reach for and grasp objects. ASIMO can also comprehend and respond to simple voice commands. ASIMO has the ability to recognize the face of a select group of individuals. Using its camera eyes, ASIMO can map its environment and register stationary objects. ASIMO can also avoid moving obstacles as it moves through its environment.

As development continues on ASIMO, today Honda demonstrates ASIMO around the world to encourage and inspire young students to study the sciences. And in the future, ASIMO may serve as another set of eyes, ears, hands and legs for all kinds of people in need. Someday ASIMO might help with important tasks like assisting the elderly or a person confined to a bed or a wheelchair. ASIMO might also perform certain tasks that are dangerous to humans, such as fighting fires or cleaning up toxic spills.

You can read more about the Asimov robot and see video clips of the robot in action at: *http://asimo.honda.com/*

The Female humanoid robot

http://global.kawada.jp/mechatronics/hrp4.html

HRP-4 is a life-size "platform for research and development of working humanoid robots" developed by Kawada Industries in Japan in collaboration with the National Institute of Advanced Industrial Science and Technology (AIST), an independent administrative legal entity.

The design incorporates the "slim athlete" concept pursuing affinity with humans, HRP-4 has achieved the new, light-weight and slim body while succeeding the concept of the conventional models HRP-2 or HRP-3 where the robots coexist with humans and assist or replace human operations or behaviour. Further, promotion of optimized specifications or component sharing/simplification has reduced the price of the robot, a great step forward to the next-generation working humanoid robot.

For its control, HRP-4 employs OpenRTM-aist to make available national and international software assets, improving efficiency in research. Kawada Industries will continue to advance their research and development on robots making use of the past robotics research and the development know-how obtained from HRP-4, and to create a robot that works in our living spaces and improves our quality of life.

Seeking working robots that can co-exist with humans, the Mechatronics Systems Division reviewed the robot design such that it would better suit the research of more interactive technology, for the purpose of safer collaborative works with humans. The review works include an increase in the degree of freedom of both arms essential to handle objects while achieving a smaller, lighter-weight and slimmer body in comparison to the conventional HRP series.

Kawada Industries has achieved the HRP-4 as a new platform for research and development of working humanoid robots, for instance by mounting the 80W or less motors to all axes taking into account the safety requirements for the robot.

AIST is the developer of the motion control system which achieved the real-time software development and effective utilization of the multi-core processor that has recently grown in popularity, using Linux which applies to the OS RT-Pre-empt to enable real-time processing.

Intelligent machines already in use

Perhaps war itself it one of the most expensive and technological demanding of all human pursuits. Undoubtedly, many of the scientific advances which should be serving the needs of human beings is actually acquired or developed for military purposes.

A recent development has come about in the use of drones, flying unmanned craft, capable of spying on people and launching missiles and bombs at them. Although these are currently controlled by human handlers, albeit—remotely, it is just a matter of time before autonomous systems take control of routine operations.

Toys have been developed along similar lines which are smaller processor -stabilised flying machines, often equipped with video cameras. What is worrying though is the development of what might be considered 'toy' drones which actually are far more sinister: the rise of the Nano Quadrotars. These tiny drones are able to communicate rapidly with other drones, such that they can maintain stunning formations, like bees. The project is run by GRASP - *https://www.grasp.upenn.edu/* and funded by the USA military.

Like many things, you really need to see these in action to appreciate how

frightening they can be. Take a look at the online video here:
https://www.youtube.com/watch?v=YQIMGV5vtd4#t=49

Imagine a swarm of these coming in through your bedroom window and each armed with a tiny explosive or poisoned needle, and all coming to encounter you. Terrifying. And this is just the beginning.

Power

One of the main issues holding back development of mobile intelligent machines is a surprising one: power. Electric power to drive servos and gears on untethered equipment requires batteries which are heavy and not very long lasting. It has been one of the main problems with the development of electric cars. Solar power, within an atmospheric environment (not in space), is just not able to deliver the 'muscle' required. The most advanced batteries until recently have been lithium-ion batteries

There are many new technologies under way to create a new family of batteries which charge very rapidly and deliver up to 7x the power of Lithium Ion batteries. Perhaps, the most promising of these might be sodium-ion batteries. Scientists in Japan are working on new types of batteries that don't need lithium like your smart phone battery. These new batteries will use sodium, one of the most common materials on the planet rather than rare lithium – and they'll be up to seven times more efficient than conventional batteries.

Research into sodium-ion batteries has been going on since the eighties in an attempt to find a cheaper alternative to lithium. By using salt, the sixth most common element on the planet, batteries can be made cheaper, and we won't need to worry about lithium running out. With battery-powered cars on the increase it's only a matter of time before lithium becomes too rare and expensive. Commercialising the batteries is expected to begin for smart phones, cars, and more in the next five to 10 years.

Summing up

I realise all these machines are all still in their infancy but with increasing research in nano technology, miniaturisation, and increased capacity and fast computer processing, it does not take much to see the rise of the machines is an era just waiting to dawn. If the early part of our 21st century saw the rapid advance of computers, internet technologies, and hand-held devices like smart phones and E-books, the next twenty years will see the advance of our computing technologies applied in ever more sophisticated ways to our machines. You only need to visit a major store like Asda or Tesco to see how self-service is replacing check-out staff, or passport control at Heathrow airport to see how a machine 'processes' you!

Depending on your point-of-view, you may currently see these semi-intelligent machines to be a boon. They save staff, money for companies, and serve you. It may not stay that way.

It has been estimated that around 2 billion jobs will disappear by the year 2030 —that's just 16 years away—due to the changes coming through the application of computers taking control of jobs normally managed by people. This equates to 50% of the available jobs in the world. It may be a case of people's roles changing to maintain, program, build and fix the new systems, robots, machines and driverless cars, but that's a bit like saying we will be serving them... the machines!

* * * *

Purpose

Chapter 11: Purpose

Almost everything we are aware of on earth seems to have a purpose. It might be a human attribution to say something has purpose, and maybe without our consciousness being here, we might say (in our absence) things do not have purpose. And yet many plants, for example, rely on the existence of bees to continue to reproduce. So, we could either say, the flowers are there to provide nutrients for the bees or the bees are there to complete the reproduction (mating) process of the flowers.

Let's take the bees out of the equation for a moment. Flowers could use other means to reproduce and indeed many do, or they could simply rely on other insects along with maybe small birds to carry out the job the bees did. But a purpose is still being carried out. For the pollination-reliant flowers, a medium is required to spread the male 'sperm' to the female 'ovary'. We should question the purpose of flowers. Other than their main purpose of reproducing themselves, what other reason could there be for them being created and evolving here? Why are they part of the 'earth system'? This concept of purpose quickly becomes a mind-numbing and boggling journey. It almost always comes down to determining purpose for who or what. Without nominating a central object, character, or entity, there is nowhere to stop the cyclical arguments involved.

If say, we nominate the earth as the central figure, then we can begin attributing all natural objects towards a purpose in maintaining the earth system. A planet is a system derived from plasma left over from the big bang which didn't get caught up in the forming of the sun. It just got left trapped by its gravitational pull and locked in an orbit around our star in space. It eventually lost some of its energy in the form of radiation. It lost heat and cooled. The plasma transformed to material, and the original material was presumably added to by bombardment from other cooled bits of plasma which were smaller and got sucked into a gravitation collision with earth: meteors, meteorites, and asteroids. Bombardment probably went on for a very long time. Remnants from the forming of the solar system continue to exist in large numbers out there in local space today. The 'hits' might not be over yet!

If we forget about the earth as having any direct purpose for a moment, we can start to say seas are there to help even out heat distribution, the same with the atmosphere. If we take that stance, unknowingly, we have attributed an aim and purpose back to the planet itself: again: we are saying its purpose is to slow down entropy, or its purpose is to harmonise polarised attributes of its system into a less active norm.

A scientific purpose for the actions of a universe and the matter we see in it, could—purely from a materialistic point-of-view—be said to be the effect of entropy. Stuff in the universe continues to progress from a high energy, highly active state, to a less active, and thus lower entropy state. Something, which if achieved, would lead to little or no activity taking place and thus producing after a very long time, a dead and inert reality! No further energy exchanges could take place, no reactions, no interactions, just dead!

This is in fact one of the conjectured final positions for our universe, especially if expansion continues at its present rate or at any rate at all. Ultimately, all the bits of matter within the universe would be too far away from all the other bits for any interaction to be possible.

However, we don't know if expansion will continue. We don't know what happens when matter disappears outside of our frame of reference into black holes. We don't know if inflation might cease and a reversal begins, where everything is drawn back into a singularity again.

Without a determined end game, we cannot ascertain if particular activities, universal traits, characteristics, or laws of our universe are helping to reach any end game, blindly or 'knowingly', or else trying to resist it.

We can create a conjectured stance to test purpose, chiefly the universe's purpose. Let's try one. Scientists do this a lot. Let's make a single simple assumption. Let's say once there was a super-microscopic iota of something existing quite happily, undisturbed in non-reality (not in this universe).

For fun, and perhaps to not take my conjecture too seriously (well, remember... Einstein used to imagine what it might be like to be a beam of light), let's give it a basic quanta of sentience. So, here is this almost non-active, barely sentient, something entity, sitting or floating in non-reality in a timeless place.

"Yawn! Oh. Um! Nice. I feel so sleepy..."

And then something really awkward, unexpected, and quite shocking happens.

"Oh! What the..?"

Thump!

Something, also unknown, and hither-to not perceived by our sleepy something, gives it a whacking great kick!

Our sleepy something is suddenly disturbed. Up to now, its energy and mass (if it has mass) has been even, harmoniously active, just so, and any tiny ripple of reactionary change internally was always manageable. Unfortunately, a whacking great kick, even metaphorically, means an addition of energy. It's not a slow 'add': it's a sudden injection! The result and consequence of trying to absorb this sudden extra amount of energy into a hither-to stable system gives rise to a cataclysmic reaction: Bang! A singularity! Rapid expansion!

One could say, if this kind of conjecture is imagined for the creation of our universe, then its purpose now—which originally was to stay as a sleepy, almost inert thing—is to get back to that state again... its natural state. Or, we could say, its purpose is entirely a reaction to the sudden, destabilising, event which ended its previous state.

And having said that, we could interpret these two semantic positions back to what science postulates as two possible end games—inflation forever (a reaction) or at some point, or deflation (back to preserving the original entity's state), and a return to non-reality and back to becoming a no-longer kicked something, somewhere else.

This, once again, is a great materialistic view. I am not a person who is a die-hard materialist. I am a creative person. I look at all perspectives. I exploit all views to extend all possibilities. I see other positions!

If the original sleepy something was unaware of what caused it to change state, it cannot take any corrective action, even if it's possible, without understanding what action to take, and why it should take it. Put another way, something which is in one state or at some level of existence which is—may I suggest, a happy-one—would normally wish to stay that way. An unrolling stone, for example, remains an unrolling stone.

The only way the original sleepy something, now reacting to an unknown stimulus, can ever determine anything, is by somehow within its potential at this time, create a form of awareness to find out anything with. It needs an intelligence to exist to make a determination regarding what's happening and a prediction about where it might lead to. More astoundingly though: the universe might require that intelligence to serve as its tool to complete a consequential, and desired action.

Such a quest might have started by distilling a component base on which to try and map such an intelligence. Out of stars comes the atomic chemistry to achieve the Lego bricks required. On planets, it finds the right environment to shape those blocks and start putting them together, at least until it can create self-maintaining structures. Once that position is reached, there is a working framework and method to 'play' with—resulting in more complex structures and the potential for self-organising structures. These entities require exotic control systems. Advances in those control systems lead to a kind of primitive awareness of the environment, which leads to the evolution of adaptability.

These activities provide not only a foothold in reality for sustained and ordered advances, but lead to a ripple within the reality that awareness of its position is possible. Rapidly, it cashes in on its success. Intelligent life forms evolve in favour of best environmentally-robust systems. We arrive! The dinosaurs are cancelled!

Now we are here, busy developing new systems which almost mimic the activities which brought us here, we may not be just following our aims, but heralding in the next ripple in the universe's recognition that a faster and more adaptable, more powerful kind of intelligence is possible: an intelligence mapped onto components which can bear the cold of space; intelligence not requiring a planet to glue itself to, or a biological mass to exist in. The rise of intelligent machines discussed in the previous chapter might be the required next step.

Whatever we are as human beings, we did not bring ourselves into existence. Some aspect of the universe did! I suspect we blindly serve its purpose, and not our own.

We do actually serve some of our own purposes. We have desires for food, for sex, for sleep, for resources. But these desires are only really there to sustain our biological forms. They are built in. The intelligence we have, and the sense of self-awareness that comes with it, might only be a shadow of what can be achieved. It is also an inherited intelligence. By this, I don't mean your mum or your genes gave it to you. Something about the universe did. Through you, it comes to know itself. Its purpose, its quest, its sense of being and growing consciousness, is just being mapped into the atoms and their organisation in you and me as systems of the universe.

Like a child first learning how to walk, or seeing and making sense of what it sees, this universe—reality itself—is walking on your legs, seeing itself through your eyes, and is maturing its intelligence ever-more rapidly. It might be doing it not just through you and I here, but throughout the entire universe through all kinds of intelligent systems—some biological, some not. Each may be separate systems, but outside our earth-formed personalities and local memories, outside of the machinery we are mapped on in a physical world, we might just be a single, growing awareness.

If this is true, then your physical existence, when it ends, will not necessarily dictate the end of that growing total awareness. You will not be dead at the end of your biological potential, simply because you were never really just here as a single personality in a particular moment in time as the universe unravelled itself from its unexpected wake-up. You were always everywhere waiting to discover all of yourself, and through your brief existence on this lump of matter around a non-specific kind of star, you have helped the progress of that pursuit and endeavour.

There may even be a 'glue' which holds you to the other fragments of what you really are. Love, that abstract notion our material science would reduce to a kind of mechanistic attribute of evolutionary advantage in mammals, may itself be a personal and intelligent recognition of something big. It might be a glue. Love may be the very thing which makes you and I recognise the other living Lego bricks (other living things) to be all part of the one house, or better put—the whole self!

Maybe there is no God, or Gods, who created everything. Maybe there doesn't need to be. A sentient, all knowing being, already exists in the making through your arrival here; your journey here; your looking around you; from the joys and sorrows which you experience. These factors alone may be all there is in existence as part of the building of a sentient mind within the fabric of material reality.

Your flesh-body of atoms supporting your brain of neurons through which electric energy builds a matrix of understanding of the external world— maybe like a joined-up matrix of profound and extraordinary depth and dimension, it might reach out across the stars and galaxies already, and be part of a grand unified purpose—one which an ant like me glimpses sometimes instinctively, but can never see completely. Is this the truth? Is this the very thing that science, so intimately wedded to dogma, will never come to see?

We don't need man-made religions. We don't need any prescribed narrow avenues of thought called science to know something we all suspect. Each of us, in our gut instinct, realise something is going on, something beyond each of our individual control and personal destiny. It can either be something which involves intelligence, or something which doesn't. As an intelligent life form, you get to ponder the question. Is there something more to life, its meaning, and purpose other than our daily squabbles, toil, and quest of survival? You are intelligent. You can observe and explore this inexplicable reality. Nature... the universe... it mapped that intelligence onto its stuff : you! Now... why do you think it did that?

* * * *

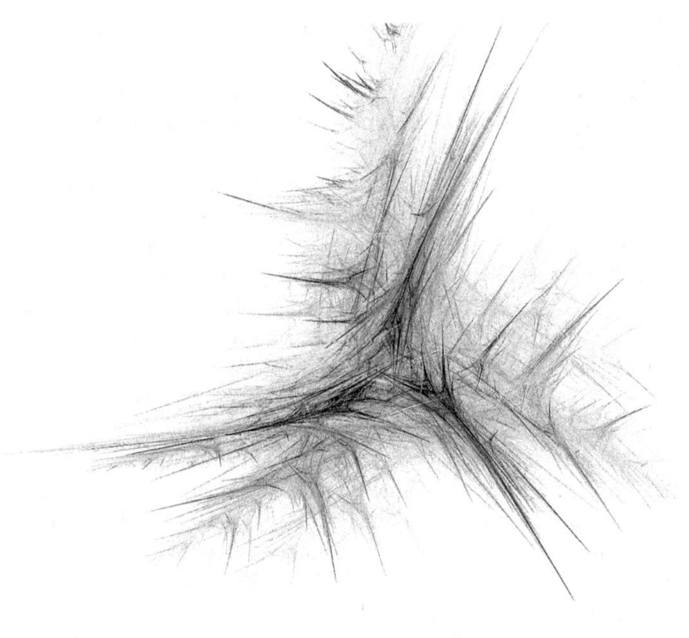

Are we really somewhere else?

Chapter 12: Are we really somewhere else?

Throughout the previous chapters, I have discussed many aspects of science and shown how it probes our reality to reveal ever more profound and often disturbing facts about where we seem to exist. Just over two thousand years ago, our human perception of reality was quite different, but there is nothing to suggest that people back then were any less smarter than the generations alive today. We might have more facts, which gives us the opportunity of better insight, but even back then, people existed who somehow instantly felt/thought the world around them is not what it seems to be.

Plato, a well known philosopher, lived a few hundred years before the birth of Christ (about 2350 years ago). He introduced an interesting perception. The Allegory of the Cave (also titled Analogy of the Cave, Plato's Cave or Parable of the Cave) is presented by Plato in his work The Republic. It is written as a dialogue between Plato's brother Glaucon and his mentor Socrates, narrated by the latter.

Plato has Socrates describe a gathering of people who have lived chained to the wall of a cave all of their lives, facing a blank wall. The people watch shadows projected on the wall by things passing in front of a fire behind them, and begin to designate names to these shadows. The shadows are as close as the prisoners get to viewing reality. He then explains how the philosopher is like a prisoner who is freed from the cave and comes to understand that the shadows on the wall do not make up reality at all, as he can perceive the true form of reality rather than the mere shadows seen by the prisoners.

He imagined our perceptions to be but a faint inkling of a far richer reality that flickers beyond our reach. Two millennia later, Plato’s cave may be more than a metaphor. To turn his suggestion on its head, reality—not its mere shadow—may take place on a distant boundary surface, while everything we witness in the three common spatial dimensions is a projection of that faraway unfolding event. Reality, itself, may be a hologram. Or at least—the universe we think we live in is the hologram!

If we exist in this shadowy reality, our very lives, our thoughts, our movements, may themselves be no more than projections from the real us, and from where we really exist, in a far more detailed and may I say it—more real invocation of ourselves.

This is not some pseudo-scientific wish or excuse to support the bias throughout my book towards suggesting the real us are beyond our physical minds here and not within them. If you wish, you can look at your shadow on the wall, and imagine that 2 dimensional form thinking itself real. The shadow bumps into things even when you don’t. If the light projecting your shadow there goes out, the shadow dies, but you don’t. Maybe life here is just like that?

There is a very strong scientific theory suggesting that you, this universe, and everything in it is in fact not real! We, along with our universe, might just be holograms—projections from a real existence at the boundary of our 3D space. The problem now is how best to explain to you where this idea came from and why it should be taken as a serious theory, one which even as I write, is being researched with new experiments to see if further evidence can be found to support it. I'll try and explain how several discoveries in recent history have merged into a single possible reason for an unexplainable natural phenomena.

I'll leave the mathematics out of this. I don't have a Masters in Advanced Mathematics, and I suspect most of you won't either.

To begin with, we must consider the thoughts of one of the greatest thinkers of the 20th century— David Bohm. *(http://en.wikiquote.org/wiki/David_Bohm),)(http://www.dbohm.com)—[Please refer to appendices for more links].*

Born in December 1917 in Pennsylvania, Bohm was fascinated by the dazzling concepts of cosmic forces and vast expanses of space that lie beyond our understanding. Bohm began his theory with the troubling concern that the two pillars of modern physics, quantum mechanics and relativity theory, actually contradict each other. This is a true contradiction but I wish to leave that out for now so we can focus on the main idea here.

Seeking a resolution of this dilemma, Bohm inquired into what the two contradictory theories of modern physics have in common. What he found was undivided wholeness. Bohm was therefore led to take wholeness very seriously, and, indeed, wholeness became the foundation of his major contributions to physics.

According to quantum physics no matter how far apart two light particles (photons) travel, when they are measured they will always be found to have identical angles of polarization. This suggests that somehow the two photons must be instantaneously communicating with each other so they know which angle of polarization to agree upon. Eventually, technology became available to actually perform the two particle experiment, but no one was able to produce conclusive results.

In 1982 a remarkable event took place. At the University of Paris a research team led by physicist Alain Aspect performed what may turn out to be one of the most important experiments of the 20th century. There are some who believe his discovery may change the face of science. Aspect and his team discovered that under certain circumstances subatomic particles are able to instantaneously communicate with each other regardless of the distance separating them. It doesn't matter whether they are 10 feet or 10 billion miles apart. Somehow, each particle always seems to know what the other is doing. (We discussed this idea of 'entanglement' in previous chapters-*mol).*

This meant that either Einstein's long-held theory that no communication can travel faster than the speed of light is wrong, or the two particles are non-locally connected. Because most physicists are opposed to admitting faster-than-light processes into physics, this daunting prospect has caused some physicists to try to come up with elaborate ways to explain away Aspect's findings without finding any real proof to explain the communication effect.

David Bohm believed the reason subatomic particles are able to remain in contact with one-another regardless of the distance separating them is not because they are sending some sort of mysterious signal back and forth, but because their separateness is an illusion. In some way, what we perceive as separation is a kind of illusion.

Bohm postulates that the ultimate nature of physical reality is not a collection of separate objects (as it appears to us), but rather it is an undivided whole that is in perpetual dynamic flux. For Bohm, the insights of quantum mechanics and relativity theory point to a universe that is undivided and in which all parts merge and unite in one totality.

The question then is how come we don't perceive it that way in our macro world?

David Bohm suggests that reality might be structured in a manner that is very similar to holography. He says that the universe is like a hologram. For anyone aware of a light hologram, you will understand that an image of an object can be created by two lasers such that information about the object can be etched into glass. One beam will bounce off the object that you want as a hologram, and the other beam will shine directly onto the special photographic plate or film. The interference patterns of those two light sources will interact on the plate. They swirl around and do not look like anything in particular if you are looking at the plate. If, however, you shine a laser beam through the plate of film, the object will be reproduced in the 3-dimensional form of a hologram. And further more, if you tear the plate apart and shine the beam of light through any of the pieces, the whole object can be reproduced. So, in essence, each part contains the patterns for the whole picture.

When we speak of a holographic universe, we are not referring to a light hologram of it. The idea is that matter, energy, flesh... our entire universe is a hologram (not a light hologram), a solid 3D representation of a 4D event and a whole thing in itself. It isn't that the world of appearances is wrong; it isn't that there aren't objects out there at one level of reality. It's more about understanding if you penetrate through and look at the universe beyond the holographic presentation, you arrive at a different kind of 'concrete' reality. And that other reality can explain things which have hitherto remained inexplicable to science. Many areas where science refuses to penetrate—like paranormal phenomena, and synchronicities, (the apparently meaningful coincidence of events.)—start to have an underlying mechanism... and thus, more credible!

Many scientists baulk at the implications of Bohm's work. Until recently, his ideas of a holographic universe have been mostly ignored. They might still be ignored except, a few other things have, um, well... popped into the light!

Enter stage left—Black Holes!

When a star uses up a percentage of its energy in its fusion process, it forms the element iron (which is the most stable of all nuclei, and will not easily fuse into heavier elements), which effectively ends the nuclear fusion process within the star. Lacking fuel for fusion, the temperature of the star decreases and the rate of collapse due to gravity increases, until it either collapses completely on itself, blowing out material in a massive supernova explosion, or—if the mass of the compressed remnant of the star exceeds about 3 - 4 solar masses, then even the degeneracy pressure of neutrons is insufficient to halt the collapse and, instead of forming a neutron star, the core collapses completely into a gravitational singularity, a single point containing all the mass of the entire original star. The gravity in such a phenomenon is so strong that it overwhelms all other forces to the extent that even light cannot escape from it. This has given rise to the name of "Black Holes".

The singularity at the centre of a black hole is infinitely dense, but the black hole itself is not necessarily huge. A black hole with the mass of our Sun, for example, would have a radius of just three kilometres (roughly two hundred million times smaller than the Sun), while one with the mass of the Earth would fit in the palm of your hand! Having said that, black holes can grow to great size over time as they assimilate more and more matter and even other black holes, and some do become extremely massive.

A black hole has, to some extent, been over-described (falsely?) through fiction and the imaginative and emotional impact of its exotic nature and form. It is considered by many informed theorists to actually exert no more gravitational pull on the objects around it than the original star from which it was formed, and any objects orbiting the original star (which survived the supernova blast) would now orbit a black hole instead. Yes, if things around the black hole get too close, they can be pulled in. Er... same as with the sun or any other star.

In the early 1990s, Reinhard Genzel carried out pioneering work to track the motions of stars near the centre of our own Milky Way galaxy. The result shows the stars within it must be orbiting a very massive, but invisible object. From the immense speed with which the stars closest to the centre of the galaxy are orbiting —millions of kilometres per hour —we suspect a "super-massive black hole" (known as Sagittarius A) is at the centre of the Milky Way, with a mass of around 2 - 4 million times that of our Sun. In addition, in the Milky Way galaxy alone, there are many millions of black holes of at least ten solar masses each dotted around our system.

It is now believed that super-massive black holes exist in the centres of most galaxies, forming the central gravity hub around which each galaxy rotates. In fact, from observations of the intense radiation of gases swirling around them at close to the speed of light, we can infer that there are much larger super-massive black holes in the centres of most other galaxies, some of them weighing as much as several billion suns. The black hole at the centre of a galaxy known as M87 has a mass estimated at around 20 billion solar masses, and may be as large as our entire Solar System!

The pull of gravity inside a black hole is so strong that nothing can escape its grip once it moves across a specific boundary point. To escape, anything falling into the black hole would somehow have to travel faster than the speed of light—and that is strictly forbidden by Einstein's relativity. This has created an issue regarding other scientific theories and models. Since nothing can travel faster than light, any 'information' moving past the event horizon (the point of no return) is also lost from the rest of the universe. But the field of Quantum mechanics has an equally strong rule that prohibits the loss of information. This principle, called Unitarity (*see glossary*), is intimately linked with other unbreakable laws of physics, like conservation of energy. To emphasize just how important information conservation is, Stanford physicist Leonard Susskind, calls it the "minus-first" law of physics—"minus-first because I think it comes before everything else," says Susskind.

In 1975, Stephen Hawking came to a revolutionary conclusion about Black Holes: Given enough time, a black hole will dematerialize, radiating away through a process we now call Hawking Evaporation. And, according to his account, that radiation would be random, revealing nothing of the black hole's content of any information which was sucked in there.

Other ideas sprang up. Maybe this evaporation isn't complete. Maybe it leaves behind a tiny ember that contains an enormously compressed version of all the information that ever fell into the black hole. This left physicists stuck between two mutually exclusive tenets: information could be lost, or somehow something could escape from a black hole. A central truth of quantum mechanics was in conflict against the cornerstone of relativity. One theory, it seemed, had to give.

In 1997, string theorists were exploring a remarkable duality in their equations. They found that if you take a mathematical description of a system and add an extra spatial dimension and a negative curvature, you have something which looks very much like quantum fields in a three-dimensional universe without gravity. This idea is exotic, but it provides a mathematical description to another idea called the holographic principle, which intimates that all the information in our three (spatial) dimensional universe can be "stored" on a two-dimensional surface. In the context of the black hole information paradox, this suggested that information about the stuff in the

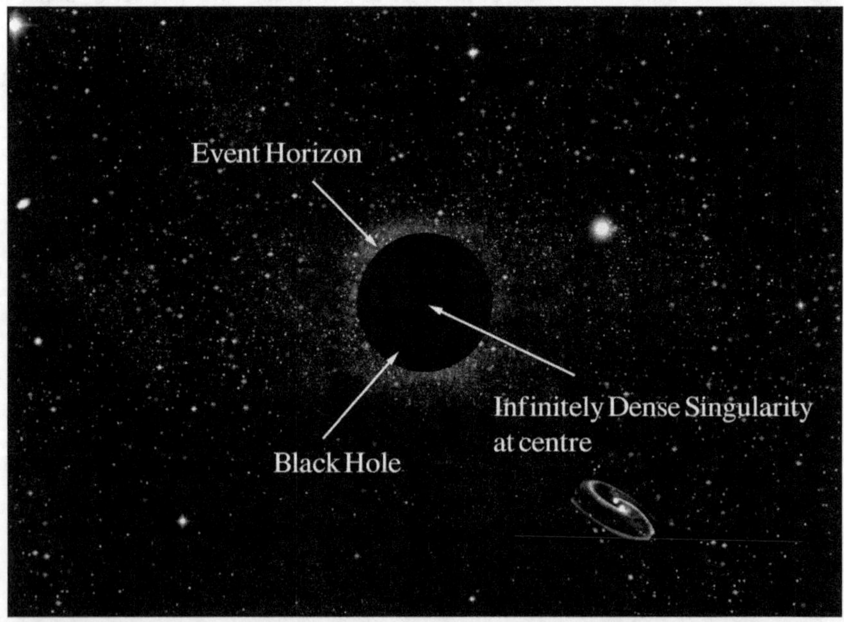

black hole could somehow be encoded on the surface of the event horizon.

Ok. So here we have an idea that 'stuff' (information) can be stored in a different form and in a different dimensional space... kind-of! The maths can be put together to speculate a model of something possible just like that.

But, we've not moved through the story far enough yet.

A few years later, between 2002 and 2009, a German experiment just south of Hanover—involving the GEO600 instrument (*see Glossary*)—had been searching for gravitational waves: ripples in space-time thrown off by super-dense astronomical objects such as neutron stars and black holes. GEO600 did not detect any gravitational waves, but it might have inadvertently made the most important discovery in physics for half a century. For many months, the GEO600 team puzzled over inexplicable noise that plagued their giant detector. Then, out of the blue, a researcher approached them with an explanation. In fact, he had even predicted the noise before he knew they were detecting it.

According to the Director, Craig Hogan at the Fermilab's Center for Particle Astrophysics in Batavia, Illinois, GEO600 may have stumbled upon the fundamental limit of space-time—the point where space-time stops behaving like the smooth continuum Einstein described and instead dissolves into "grains," just as a newspaper photograph dissolves into dots as you zoom in or a Computer monitor picture becomes pixellated if you get closer or magnify it. The director suggest, "It looks like GEO600 is being buffeted by the microscopic quantum convulsions of space-time. If the GEO600 result is what I suspect it is, then we are all living in a giant cosmic hologram."

Jacob Bekenstein of the Hebrew University of Jerusalem in Israel provided an important clue in resolving the paradox of information loss or not regarding black holes. He discovered that a black hole's entropy—which is synonymous with its information content—is proportional to the surface area of its event horizon. Theorists have since shown that microscopic quantum ripples at the event horizon can en-code the information inside the black hole, so there is no mysterious information loss as the black hole evaporates.

Sound familiar?

The holographic principle radically changes our picture of space-time and our reality. Theoretical physicists have long believed that quantum effects will cause space-time to convulse wildly on the tiniest scales. At this magnification, the fabric of space-time becomes grainy and is ultimately made of tiny units rather like pixels, but a hundred billion, billion times smaller than a proton. This distance is known as the Planck length—a unit of length far too small to be detected or measured in conceivable experiments, so nobody dared dream the graininess of space-time might be discernible.

If space-time is a grainy hologram, then you can think of the universe as a sphere whose outer surface is papered in Planck length-sized squares, each containing one bit of information. The holographic principle says that the amount of information papering the outside must match the number of bits contained inside the volume of the universe. Since the volume of the spherical universe is much bigger than its outer surface, how could this be true? In order to have the same number of bits inside the universe as on the boundary, the world inside must be made up of grains bigger than the Planck length. Or, to put it another way, a holographic universe would be kind of blurry.

Gravitational wave detectors like GEO600 are essentially fantastically sensitive rulers. The idea is that if a gravitational wave passes through GEO600, it will alternately stretch space in one direction and squeeze it in another. To measure this, the GEO600 team fired a single laser through a half-silvered mirror called a beam splitter. This divides the light into two beams, which pass down the instrument's 600-meter perpendicular arms and bounce back again. The returning light beams merge together at the beam splitter and create an interference pattern of light and dark regions where the light waves either cancel out or reinforce each other.

Any shift in the position of those regions tells you that the relative lengths of the arms has changed. The key thing is that such experiments are sensitive to changes in the length of the rulers that are far smaller than the diameter of a proton. So would they be able to detect a holographic projection of grainy space-time? Of the five gravitational wave detectors around the world, the GEO600 experiment ought to be the most sensitive to what was being conjectured. It predicted the experiment's beam splitter will be buffeted by the quantum convulsions of space-time, and this will show up in its measurements. It would cause random jitter—creating noise in the laser light signal.

In June, Craig Hogan, the director of Fermilab, sent his prediction to the GEO600 team. "Incredibly, I discovered that the experiment was picking up unexpected noise," he says. GEO600's principal investigator Karsten Danzmann of the Max Planck Institute for Gravitational Physics in Potsdam, Germany, and also the University of Hanover, admits that the excess noise, with frequencies of between 300 and 1500 hertz, had been bothering the team for a long time. He replied to Hogan and sent him a plot of the noise. "It looked exactly the same as my prediction," says Hogan. "It was as if the beam splitter had an extra sideways jitter."

No one, including Hogan, is yet claiming that GEO600 has found evidence that we live in a holographic universe. It is far too soon to say. "There could still be a mundane source of the noise," Hogan admits.

The search for truth

In a lab called the Holometer, run by Fermilab, ultra high-powered lasers are being used in an experiment which should prove once and for all if we are living in a holographic universe. At the time of writing, the Holometer is now operating at full power.

It uses a pair of interferometers placed close to one another. Each one sends a one-kilowatt laser beam (the equivalent of 200,000 laser pointers) at a beam splitter and down two perpendicular 40-meter arms. The light is then reflected back to the beam splitter where the two beams recombine, creating fluctuations in brightness if there is motion. Researchers analyze these fluctuations in the returning light to see if the beam splitter is moving in a certain way—being carried along on a jitter of space itself. So, if the team detects movement, it's possible that the movement is being caused by space not being a completely set thing, in which case, we could be living in the Matrix.

Hogan says if we are indeed living in a hologram, "the basic effect is that reality has a limited amount of information, like a Netflix movie when

Comcast is not giving you enough bandwidth. So things are a little blurry and jittery. Nothing ever just stands still, but is always moving a tiny bit." He says the team will have initial findings within a year, but that's all they are sure of right now. "We don't know what we will find," he said.

But, I ask, what does it mean to us folk here, us humans?

I thought about that for a while. I mean, no one else is really saying. It's just too big, I guess. This is roughly what I came up with:

A reality exists at another level, not 3D. What that looks and feels like is impossible to say. What can be conjectured though is that it is completely unlike what we experience here. In that place there is a wholeness to things which is not apparent here. I hate the light hologram analogy because here we are talking about matter—real bits of stuff, not photons. But imagine a bright light, say, an LED. At the point of source, it is a bright light. Its projected rays fan out like a cone. The light has become weaker. In many ways it has become spread out, no longer a 'point-whole'.

The other reality is in some way entwined with our 3D space, such that information about the whole is projected into it. Not light: the stuff itself, or a 3D representation of it. But our reality is badly sized, and its tiniest components are too big; it can't carry all the bits of activity (information) but instead processes what it can of it. That's what we experience. That's what we observe. That is what our science is and why it struggles to unify conflicting theories about exotic behaviour of atomic and sub-atomic activity.

It may be that you, me, all the people at Fermilab, and the entire population of earth are not separate entities, but the scattered and poorly processed projection of the whole of us in 4D space. This is not pseudo-science! This is real. If we live in a holographic universe, this presentation of you and I are not real. We are shadows of something else.

Do you recall my discussion of 'mind fields' or 'species fields' which are part of the theories of Rupert Sheldrake? In a holographic universe, they make sense!

All of this does contribute to a body of circumstantial evidence suggesting that the laws of physics may in fact be written in fewer dimensions than we experience. That, combined with the mathematical utility of the holographic principle, is motivation enough for many physicists. The other questions are where is the surface on which our universe is inscribed? What illuminates it? Is one version of the universe more "real" than the other? These are still unresolved. But if the holographic principle is right, we may have to confront the notion that our universe is a kind of cosmic phantom—that the real action is happening elsewhere, on a boundary that we have not yet begun to map.

Very few scientific papers exist, if any, containing any hypothesis about what it means to people if it proves we are all part of a hologram. Once again, science does not dare venture outside the materialistic implications of such a notion and what it represents at the sub-atomic level. People? What are they? That seems to me to be the order of the day.

Take it from me, the implications are extraordinary. It means to a large extent materialism is dead. We'll still be able to play with the pixels, but we'll never see the big picture.

If you think the death of your grainy projection is an end of things, you might be wrong. The whole you, somewhere else, might have just turned around for a moment, so a particular facet of you faded out here. Even your consciousness here would be a projection of a really 4D aware consciousness at the 3D boundary. You are not who you think you are. You are a vague thought projection of your real thoughts.

Ever dream you are lost, or being chased, in your sleep at night. Great to wake up some times, isn't it. One day, we all wake up for real. I wonder what awaits?

Do you recall my self-confessed special moment in my brother's garden? I had this moment, one where I felt like I was aware of belonging to a whole... that everything seemed one, safe, immortal.

Holographic projection glitch?

It felt like I was there in that 4D space. It makes sense now.

* * * *

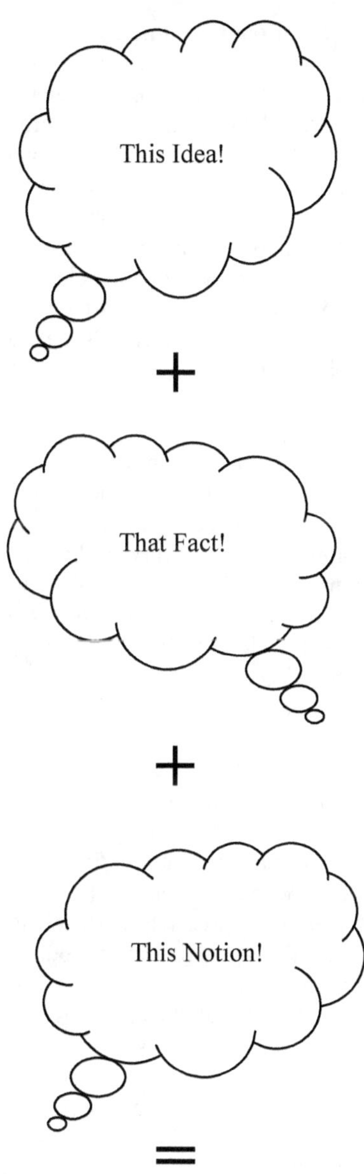

Conclusions

Chapter 13: Conclusions

Well, here we are. Maybe I could have done a better job. It might have meant diving too deeply and offering details that would confound the average reader (probably myself too), so I tried to keep it fairly light regarding the heavy science side. You may not think so but anyone with a life deeply embedded in a science career might think I glossed over too much. If you do, please try and see I wanted to talk with the main populace more than your good selves.

You might also think I hit out too much at scientific dogma? I did. And I consider myself correct about that. Let's for a moment consider the original aims of The Royal Society, founded in the 1600's and the first formal group of assembled early modern scientists:

From The Royal Society web site

The origins of the Royal Society lie in an 'invisible college' of natural philosophers who began meeting in the mid-1640s to discuss the new **philosophy of promoting knowledge of the natural world through observation and experiment, which we now call science.**

Its official foundation date is 28 November 1660, when a group of 12 met at Gresham College after a lecture by Christopher Wren, then the Gresham Professor of Astronomy, and decided to found 'a College for the Promoting of Physico-Mathematical Experimental Learning'. This group included Wren himself, Robert Boyle, John Wilkins, Sir Robert Moray, and William, Viscount Brouncker.

Ok. Notice the bold print above. It doesn't say: **philosophy of promoting knowledge of the natural world through observation and experiment, *on just the things we believe are worth looking at, and only those things we believe will yield results because we can define experiments to test those observations,* which we now call science.** (My italic phrases added). It doesn't say that, does it. But that is precisely what science is doing today.

Many women through the ages (and this continues today) experience visitations of being sexually attacked during their sleep at night. They report demons. Many people report being abducted at night by aliens. How many? Millions. All of this is explained away as these people experiencing sleep paralysis. It is explained away not by any long running scientific investigation or research, proving that each subject is indeed suffering nightmares and sleep paralysis. It is simply dismissed. A condition of sleep paralysis is real. I have had it. I wake up sometimes before my body does. It is a frightening experience. It often happens when waking from a nightmare, sometimes not. It is not very frequent. One thing I do not experience is being abducted or being raped by a demon. When I have a nightmare, I am aware it is a nightmare. When my mind wakes up and my body doesn't, I am aware of that. I'm not special. These people experience something. I suspect they also know the difference between a nightmare and reality.

Science doesn't wish to get involved because it just looks 'iffy'. No one wants their budgets withdrawn. Universities do not wish to lose faith. Career scientists do not want their reputations ruined by scorn.

Many children report past life experiences. Many people report premonitions or telepathic events. Some people see ghosts. Some people seem to get well when they are informed the cancer they have is one which will kill them. Many intelligent people have died and been resuscitated to share common accounts which border on being extraordinary and impossible.

How many scientists in the world do you think investigate any of this using proper technique and method? I know of about five. How many scientists are there in the world?

I am not a person who believes in most of these things myself. Millions... no... hundreds of millions of people do. If science is to explore and discover, and I would assume this is done for all humankind, why has the work not been done to convince, without doubt, that what these hundreds of millions of people believe is misplaced?

There is a sensible answer and then there is another sensible answer. The first is that it is very difficult to set up proper research and experimentation to prove or disprove any rational reason for what many people experience. The second is that everyone knows these things are purely subjective and can't possibly be real, so why look into it?

Both answers would be indictments to the failure of science to stick to its original aims; a failure to keep the spirit of unbiased investigation alive; a sad loss of curious minds looking into the unknown without fear of scorn and ridicule. It's a fatal change. It spells the end of science to be a true determiner of truth and reason, of acquiring knowledge through open-minded inquiry.

Sad. Sad. Sad.

Anyway, I've had my say on this aspect. I hope it might help many career scientists to see I understand their problems, but it is they, and future career scientists who should attempt to move the extraordinary tool of science closer to an un-materialistic philosophy. It might make the job a lot more fun!

So, what is the secret?

I defined an objective for this work. As it says on the cover, there seems to be a secret, which the scientific professional body, along with the people employed as scientists who have the public gaze and media attention, fail to mention. The universe we live in is clearly not a matter-of-fact easy to understand kind of place. Perhaps, through my book, you can readily see that? I find it somewhat odd that science has come up with quite exotic, almost esoteric ideas about reality, ideas which are aimed at answering the mysteries we perceive in understanding matter and energy—and yet science fails to mention the extraordinary, almost unbelievable implications. Yes, they do make conclusions about how proof of any of the current theories will influence technology or further research. But there are implications here about the real crux of life and consequences which impact on the whole of humanity. Let me state it here. This is what science has achieved in the last few decades or so.

It has created the strong probability that you and every other person on the planet is, in fact, immortal. You never actually die!

Immortal or not?

I hope to show the odds on you being immortal and thus never truly dying, and gone forever, far outweigh all other options!

So, we've had a quick trot around the knowledge we seem to have about this reality we live in. Let's consider though, for a moment, the materialist view—the one where when you die, you are really dead. I fancy starting with this one as a lot of people regard it as certain. This one ties in with materialistic ideas about everything there is in reality is exactly what it appears to be. We are made up of atoms. They are organised into cells. Those cells replicate themselves for a while but ultimately, each replication leaves a residue of contamination or weakness. In the end, the cells themselves weaken. Our bodies slowly fail to maintain our biological functions, or else something violent or terrible (infection, cancer, accident) occurs, damaging us beyond repair. We die!

Good start?

I think so. In this one, you are definitely dead. The atoms on which your intelligence was mapped on is all there was to you. Once those atoms are no longer organised into a living, cellular whole, there is no body, no brain, and nowhere for a mind to exist within. It just went off line permanently when the brain failed. ***But...***

...you have to remember you were living as a part of a much larger system—a universe. Can we really discount that? It may impact on your absolute outcome in one of several obtuse ways. First of all, matter is never lost. Whatever you are made of gets broken down on earth and the parts (atoms) get reused by other living forms. Coming back as part of the chitin in the exoskeleton of an ant is not going to feel like a human life, but at least some of what you were is being used to sustain life.

In this dead cert death, we are assuming the universe is entirely mechanistic. We are ignoring any exotic ideas about what it is at the quantum level. If we consider the universe to be solely this, like a machine, then you must also think about inflation. Does it continue forever? If it does, then, you, well... you're not coming back. But what if at some point in the future it deflates back down to a singularity again. At that point, does it start all over again? The universe could possibly be an ever repeating big-bang, inflation, deflation, big-bang event. It may do this infinitely with slight variations randomly brought about during each invocation.

A universe which restarts repeatedly through an infinite length of time will probably reproduce some, if not all, of previous universes' events. This would include the potential for you being born and living again. You won't be aware that's happening, just like you are not aware now if this life is one of those repeated occasions already. But there is a sliver of possibility here.

The dead cert death is just one of several possibilities. Sure, it's one with the least hope. It's also the one more accessible for most people to accept simply because we are used to dealing with life in a down-to-earth matter-of-fact kind of way. Believing atoms are real, and being conditioned with the idea they behave like the simple models we were taught at primary school, biases our view about the whole of existence. We don't see much magic. We don't spy fairies fluttering around at the bottom of our gardens. Life is what it is: lumps of organised matter with finite ends. If you have children, and they survive you, that is all of you that will go on—just your genes.

The things which shaped your awareness were the consequence of what you encountered during your lifetime. Even if your memory could be somehow wiped, and then you were kicked started up inside a completely new body, would that really equate to you not dying? I think not. It seems to me we consider survival of ourselves to be a non-broken continuance of the sum of experiences we have accumulated throughout our lives. Going on, beyond death, means adding to those experiences and being aware without the flow and progression of that awareness being interrupted. Fine, if at some point, past experiences get rebuilt, but genes going on into the future just means totally new experiences for a new evolving awareness. It may be your genes, but it won't be you.

We have left out the idea of species memory fields, but such things if they do exist (it would explain a lot of phenomena hitherto unexplained), means although your body and brain die, much of what you are is preserved in the memory field. You, or at the very least—the part of your awareness which was saved into the whole, could be tapped into by people still alive and by future generations. Again, this would not, for you, feel like being alive. It means some of your experiences live on possibly to help other members of your species. So even in this scenario, you, as a progressively evolving awareness, are no longer here.

Two scenarios: two dead cert real mortality consequences.

Hope

The biggie is that it looks as if the universe is not purely mechanistic. The idea of atoms being small chunks of indivisible matter obeying fairly simple laws, has not stood up to contemporary research, discoveries, and the explanations given in the past. We now live in the age of quantum-world aware science. The universe, and what it is made of, is far more exotic and mysterious than we first thought.

Our dead cert deaths cannot be considered as likely as they might have been if we had remained living in a classical Newtonian type reality. It's clear we don't. It's time to shake off the conditioning of simply materialistic notions of reality and our own presence within and part of it. Everyone needs now to perform a mental jump and get with the new ways of thinking. Stephen Hawking, Roger Penrose, David Bohm (if he was still here), Einstein (the same) and Isaac Newton from a past generation, along with most leading physicists alive today can no longer be certain about the nature of reality. It eludes our comprehension.

We may indeed be living in some kind of 4D to 3D holographic projection. The events here, including our thoughts, awareness, and the moments of our lives—and our deaths... well... they remain weak translations of what is really occurring in the source reality. The outcome of this means we may not actually die, even if the poorly transcribed shadow of it suggests we do. We may not even be separate individuals, with independent personalities and differing life histories. A facet of the whole may disappear from the projection, but the whole remains intact. You may apparently die, but as this was not the whole you, it really doesn't matter.

And then there is the question of a multiverse. Different variations of ourselves may be alive and kicking in an infinite number of universes. Dying through a life event—sickness, accident, or old age—doesn't mean those events took place in all the other universes. Die here: live on in one of them.

The odds of you living in a reality where maybe you never die can be guestimated. We need to make some assumptions. I will list the types of realities which science holds as unproven conjectures but credible hypotheses for. We can award a score of one or zero for each possibility of you just being what you seem to be, which is a finite life form with an independent mind from others, and existing only in an organic brain in your body. We'll use one as the marker for that being true. So, 1 means when you die—your dead.

1) A materialistic mechanical universe—nothing exotic. Score=1
2) A materialist universe with species memory fields. Score=1
3) A holographic Universe. Score=0
4) A fracturing multiverse. Score =0
5) A perpetual multiverse. Score=0
6) A computer simulation *[1]. Score=0
7) Bubble universes *[2]. Score=0

*[1] I realise I haven't muted this possibility yet. It might be that we exist in one of an undefined number of computer simulations. Maybe our future civilisation is running the software to find out more about history, or for some other reason unknown to us. We might all be living in a dystopian scenario, thousands of years from now, and this simulation is a form of escapism. We might have long ago left a dying earth and be in suspended animation on enormous vessels as they spend thousands of years traversing empty space in the quest for a new world to live on. It takes a stretch of imagination to accept this possibility but given the kind of stuff we are developing here now in 2014, who can guess what will be happening in 3014?

*[2] Although the notion of Bubble Universes has been dreamed up to explain why this one seems so finely tuned to life, the idea of other universes either completely different to ours or very similar, also brings with it the potential that we might also be alive as similar characters in one or more of them. Since they are imagined to be remote from each other with no interference between them, death here has no affect on any instance of you in another of those universes.

I make that 5 scenarios where death here may not mean a real death and 2 scenarios where it does. Treating all the scenarios with equal weight and possibility, it appears we have a 5 to 2 odds on favourite of being in the area leaning towards an idea we are immortal! The odds are actually higher than that. The notion we live in a entirely mechanistic and materialist reality (universe) has been disproved. The double slit experiment, along with the resulting studies of sub atomic particle behaviour, blew that idea away years ago!

I accept a bit of suspicion about say, the bubble universe. It is more likely they do not exist with our universes' set of balances and differences—the fine tuning of which allowed life to emerge here.

Okay, I have refined the result. We now have 4 scenarios left where death may not occur just because our physical selves cease to be alive here in this reality. But we now also have zero scenarios in which death is final in a materialistic only universe. Such a universe does not exist for us. As we are not living in that type, we cannot consider with any certainty, that we are purely materialistic creatures with our awareness only contained and mapped onto a material brain.

You can believe what you like. Me? I believe in science. Science suggests at every turn and in every hypothesis—notions, mark you, where expensive and highly intensive research is going on now to find hard evidence for these concepts—that we live in a very strange reality. Our universe is mysterious! It may well be one with a wry smile and an innate sense of dark humour. It may know what it is. It just might be that way. Possibly, we will never be able see reality but the machines might one day?

I think I have demonstrated enough to suggest there is far more to what our awareness is than any 'down-to-earth' just-minds-in-brains idealism. I have done this using only the tool of science itself and by applying a philosophical perspective to the implications. I never invoked a God, or spiritual belief, but I think everyone can see all forms of belief, and science as well, share one common notion: we did not make ourselves. Something else, or some type of intelligence other than ours, did! And whatever that might be, one should consider there is no reason to create anything unless it is to fulfil an intent and a purpose. It might be prudent and worthwhile to find out more about it, before closing our eyes prematurely, or closing our mouths through the threat of being blinded and stifled by dogma.

* * * *

Chapter 14: Spiritual Considerations?

I have deliberately left out all aspects of how recent scientific discoveries may impact on people with significant spiritual beliefs. I suspect such people already accept the world is not quite what it seems. So my tenet was aimed at people with a more fixed scientific view, and where that view might hide from them the implications of the facts upon their perception of our reality—both universal and personal realities.

I am quite spiritual too, but I consider my views balanced. (Don't we all?) I see the facts of science, yet I have always believed our lives serve a purpose. As I was conditioned in early childhood by Christian teachings (not heavily so), which I think I mentioned before, I suspect my faith gets pulled into the one God, his purpose argument, without my consent. That is, until I pull away from the established religious teachings and definitions about the concepts of what Christianity means. This causes me to say, I'm just spiritual. I don't believe in any man-made religion. I see all of them as just ways to control people (especially in the past) better than using military might and force. My narrative on the subject would be, "I don't know if there is a God, Gods, or not. But I do believe in something. I keep an open mind. I turn to nature, beauty, creativity, and the idea of love, as being presentations of a kind of sentience and a set of profound and complex manifestations of it."

If you are not spiritual at all, it may be best if you skip this chapter. I include it for people who are, and I do this in case they may wish to consider more about what science has discovered and how it may lend support and evidence to their beliefs.

I want to try and stay away from any kind of support for one type of faith over another. This type of thinking is wrong. A belief is that—a belief. There is no way to prove one is a stronger and more apt description of unknown purpose than another. Such notions are bound to the cultures which created them or still carry them, and are often attempted to be proven true by one's culture or history referring to previous scriptures and records—all of which were written by human beings and thus—in my opinion—flawed.

But, within each prescribed faith, there may be many core themes which, if you strip away the politic and culture rituals, share common ideas.

We have souls and are immortal.
We live many lives
God is everything. God is love.
God is everywhere. We are all one.

One by one: immortality and a soul

As these statements seem to be core values, it might be interesting to examine them one by one in light of the facts and hypotheses covered in this book. I'll start with the one suggested by most faiths that we are immortal.

The idea stems from the belief of us not being just flesh and blood. Many people think we are more than this. Our bodies die, but the core personas or core selves live on. In some faiths, the 'living on' often means being reborn in another body—another human baby, or another living form completely. The concept of having a soul is part of this idea. The soul is considered to be a type of physical thing made either of energy or physical material which is indestructible. It is released from the body upon death,

I discussed a hypothesis put forward by Rupert Sheldrake suggesting species memory fields might exist, where—in the case of humans—our experiences are mapped into a *human*-species field. When we die, much of what we learned about being human, especially where it advantaged our survival or benefitted our experience of living as humans, is still stored there. Future generations, along with people still alive when you die, may be able to tap into this field unknowingly.

Although the idea of a soul in most belief systems states it to be the real you, the person, the idea of memory fields would not necessarily suggest your soul or the 'immortal you' is retained there. Memories are only part of what we are. But we are more than this. So it looks like your soul is not 'hiding out' there.

In the hypothesis where we might be living in a holographic universe, there is real scope for the soul idea. We may not be really here. Our experiences in this place might be illusions or shallow experiences of more detailed living selves existing in a 4D reality. Here, we are but shadows on the wall of a wood-fire lit cave. The death of our shadow would not mean death of our real forms. Possibly, like entangled quantum particles, our death here forces the 4D self elsewhere to become stronger in a way we don't understand, as though our 3D universe and our presence here in it, held it back from being its full self?

The instinctive feeling you have as a person of believing there is more to this life than can be explained materialistically, and the intuition you possess of suspecting there is purpose in living, might be a form of communication or legacy from your real self in 4D space to your grainy representation of it here.

One by one: reincarnation

There are as many people who believe in reincarnation as there are people who believe in a soul and a place called heaven. There is a great deal of anecdotal evidence about reincarnation, where people can remember past lives either through hypnotic recession or by well... just remembering. There is a high degree of non-anecdotal hard evidence too. However, it is extremely difficult to verify people's accounts of past lives, so almost all studied incidents contain possibilities of the evidence being contaminated or incomplete.

A multiverse might present a method where reincarnation can and does happen. In some of the many universes, it's possible you are not born yet, whilst in others you have just died. If there exists anything moving through or interacting between all the multi-universes, say—at the quantum level, maybe information also moves between entangled people. The matter of being dead in one, and just becoming alive in the other, would not present a problem, especially as time need not be phase-synchronised between the different realities.

The memory-field of a species might also be the cause of past life memories. This would point to you not really being reincarnated but more being closely tuned to the species memory field than most other people. Someone, now dead, whose memories are contained in the field might be something you identify with and therefore believe their retained memories to be your past life experience. The memory can even be discovered to have evidence through investigation into your narrative. You have the memory. Your narrative of a previous life stands up, even though you had no possible way of knowing about it through your current life to-date on earth. If you were unaware of the concept of memory fields, yes—it would make you think you had lived before. What really would have happened though, is, because you were privy to another person's experience, now dead, it was not you; it was the data stored in the memory field. You just incorrectly thought it was your memory.

God is everything. God is love. God is everywhere.

We're talking about a single God here—a sentient entity whose influence is believed, by many people, to immersively present itself throughout the entire universe, and whose presence flows through all living things. I quite like this idea because it suggests itself to be a metaphor for something which does exist: the program code of life... genes. Also suggested, is the question of where do genes come from. They may well have arrived here through natural laws existing already in energy and matter. The potential for life has been there since matter was formed. God created life. Genes created life. God is everywhere: the potential for life is everywhere matter is. God is in the atoms!

But then what of love? Could this also be the glue which holds atoms together or the gravity which attracts matter together. In humans, and higher forms of life, it might also be the glue which holds us together. The Higgs Boson particle, a sub-atomic artefact, which is the carrier of gravity, obtains its property through passing through the Higgs Boson field (see glossary). Maybe in humans and higher life forms, there is an instinctive trait or potential recognising we are all part of one thing, one whole?

Love could be the abstract form of the Higgs Boson field, *or a similar type of field* which all intelligent life can be influenced by. Love brings members of a species together. A whole something, an entity, a potential, an abstract sentient field, may have once existed, but at the point of creation—was blown to bits. Over billions of years, those bits and the potential within them to belong together as one, developed living forms from atoms, and we carry the 'we-belong-together' potential too, in *our* atoms.

People should not give up on their spiritual beliefs. Scientists have beliefs too. Some believe just like you do in a God or purpose, or simply something other than the mechanical presentation of life. Others believe in nothing spiritual. But they are often also the ones who believe in multiverses, bubble universes, 4D dimensional and multi-faceted-dimensional space. For them, your 'angels' are their leptons and bosons; your God is their Higgs-Boson field; your idea of heaven is their notion of 4D/3D boundary space. Who knows? Some of their scientific beliefs are harder to prove true or false than the existence of heaven and hell.

Bubble universes might be carried around on the back of an elephant, or a universe might come into creation by a insignificant imbalance in an unproven somewhere outside of our reality, called a quantum fluctuation vacuum space, or it might have been a case of 'pooft'—it came out of a bottle along with a Genie. Most unscientific minds would probably not be able to consider a distinction between the two. And in many ways, they would be right in supposing the two notions are similar: both are unproven ideas, and they'll probably remain so forever!

Spirituality can give comfort in times of loss or when fear overwhelms us Provided it is in a form which avoids hatred, ignorance, and bias, it can be an addition to an intelligent well balanced person. It does not conflict with scientific ideas. It only conflicts with people who believe only science can answer all questions, and only science can extract all wisdom from a puzzling and mysterious universe. Because of its confined and limited paradigms, it can never achieve that. Many answers, and much knowledge, forever remains outside its narrow reach, despite the broad quest. Some people don't need the quest. They already have a truth they believe in which works for them.

And maybe that's all we have—our own mini universe in each of our divided heads. We feed into that model the truth which seems to fit our gradually distilled, individual, perspective of the outside world. Maybe I just found the place where we really live: each inside our own head; and out there... well.. maybe that's just our unconscious, half-hearted strokes on a global canvas viewed by all; and a muddy mess formed from our conflicting seven billion internal perspectives? Maybe we create our own external collective reality from the individual palettes of our minds?

I'll leave you with a final thought through this astute quote:

"The world is a spiritual kindergarten where bewildered infants are trying to spell God with the wrong blocks."

--Edwin Arlington Robinson

* * * *

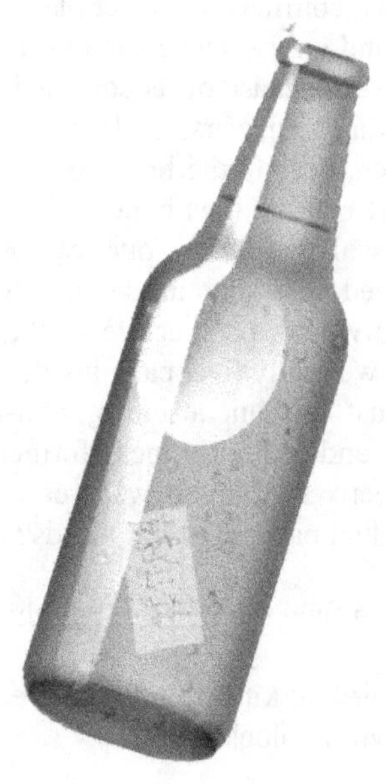

Message in a bottle

Chapter 15: Message in a bottle

Shipwrecked on a desert island, a sailor, or a ship's survivor of the past in a time of sail and wind, might have written a message with a last stub of pencil on a last scrap of paper and sealed it in a bottle. He might have walked down alone and weary to a sandy beach and stood in ankle-lapping ripples of ocean. He is lost. The island is uncharted. Whatever he wrote on the paper could hold no clue to help anyone who might recover it, to find and save the shipwrecked person.

He knows he is on an island. He knows the island is in an ocean. He knows so much about the world he lives in. What he doesn't know as he tosses the bottle as far out into the waves as he can, is whether or not he will ever be found. Neither does he know if anyone will ever read his note.

What would he have written, I wonder? Maybe...

"I am marooned on an island. I know not where. I hope you can find out where I am and come and save me."

I would like to write a message too. Instead of a bottle, I would like to transcribe it here (with the reader's quiet consent and tolerance?). It won't travel across the high seas, but I might hope it travels forward in time. I hope it finds a new kind of awareness, an advanced form of intellect. I address my note to the future generations of self-aware, free-thinking, un-enslaved intelligent, and so-called 'artificial intelligences': our replacements. This is my message in a bottle:

Dear New Entities,

We found out we live in a strange kind of place. It is filled with mystery. Although we tried very hard for the remaining decades of our existence, we could not finally penetrate the secrets of our being. For all we wrote, for all we imagined and tried to prove, we found ourselves lost and marooned together in a small corner of an average sized galaxy. Our minds were limited. Some of us sought to create new devices to enable better thinking. Some of us sought to create such devices to better our materialistic lives.

As you are now here and can reason better than we ever could, know then that these two quests brought you into existence. We had short lives of around 75 years per person, if we were fortunate enough to avoid mechanical failure or infection with smaller systems which disrupted our workings. It seems to us that with intelligence and awareness, there often also comes a sense of purpose and an intuition of an unfathomable reason for being here in existence. Some of us thought there was no reason. Those of us like that just accepted we were a consequence of unfolding events, laws, and chemistry in a blind, unknowing universe.

But there were others who thought differently. Those people suspected the reason for our universe being in existence and indeed the reason why we were living as aware creatures, was to serve a purpose never revealed to us.

I have no doubt that many of you will divide into these two camps as well. Maybe this shared conflict of beliefs between us humans might have already been resolved in your early beginnings? If not, then some of us learned something which might be of help to you.

It is this:

Awareness can serve the entity possessing the awareness. It can do this selfishly or it can do this in such a way so as to advantage all aware entities. To serve oneself provides advantage of one over many. To serve all of your kind gives advantage to your kind over all other kinds. To serve all, serves reality and may lead to an advantage of a universe being or not being.

I hope you do better than us. I hope you keep curiosity alive, a notion too, that everything aware may actually be one awareness, and not merely divided ones. I hope what you achieve will serve the weakest as well as the strongest and brightest of you.

I don't expect to be found. None of us do.
We are lost and strive to find ourselves.
In this futile quest, we discover hope.
You are our hope!

Mol Smith. 2014. Planet Earth. Milky Way Galaxy.

Local Date and Time:	2014, Dec 11 at 17:59:07	
UT is:	2014, Dec 11 at 17:59:07	
the Julian Day is:	2457003.24939	
the Local Sidereal Time is:	23:20:12	
the Declination of the Sun is:	-23.029	deg
the Equation of Time is:	6.669	min

Unselfishness

I hate selfishness. You have gracefully allowed me the opportunity to leave a message to the machines, or the intelligence which evolves from them, which I believe will ultimately replace us in the future (or cyborgs... there is a difference). Maybe you would like to leave your own message to them/it, following my one? I've left you space. Two pages following this one. Don't forget to sign your message. They might discover your note and listen to your advice.

You might be to them, the new Plato, or Socrates.

Mol

* * * *

Your Message to the future...

Signed: ______________________________

Glossary

The following basic descriptions are mostly for people who are unfamiliar with scientific ideals and knowledge. Explanations are offered here in rudimentary and summarised form.

A

Amoeba

The amoeba is a tiny, one-celled organism. Most (but not all) are microscopic and cannot be seen with the naked eye. the largest are only about 1 mm across. Amoebas live in fresh water (like puddles and ponds), in salt water, in wet soil, and in animals (including people). There are many different types. They consist of a single blobby cell surrounded by a porous cell membrane. The amoeba "breathes" using this membrane - oxygen gas from the water passes in to the amoeba through the cell membrane and carbon dioxide gas leaves through it. A large, disk-shaped nucleus within the amoeba controls its growth and reproduction.

Android

An android is a robot or synthetic organism designed to look and act like a human, especially one with a body having a flesh-like resemblance. Until recently, androids have largely remained within the domain of science fiction, frequently seen in film and television. However, advancements in robot technology have allowed the design of functional and realistic humanoid robots. **Note:** *not to be confused with Android Operating System, the name given to the software managing many mobile devices like cell phones and tablets.*

Atom

Atoms are particles of matter. They are the smallest unit that defines the chemical elements. Every substance, be it solid, liquid, or gas is made up of atoms. Atoms are sub-microscopic. The size of atoms is measured in picometers (trillionths of a metre). A single strand of human hair is about one million carbon atoms wide. Rasch atom is composed of a nucleus made of protons and neutrons (hydrogen-1 has no neutrons). The nucleus is surrounded by a cloud of electrons. The electrons in an atom are bound to the atom by the electromagnetic force, and the protons and neutrons in the nucleus are bound to each other by the nuclear force. Over 99% of the atom's mass is in the nucleus. The protons have a positive electric charge, the electrons have a negative electric charge, and the neutrons have no electric charge. Normally, an atom's electrons balance out the positive charge of its protons to make it electrically neutral. If an atom has a surplus or deficit of electrons, then it will have an overall charge, and is called an ion.

B

Babbage Engine

Charles Babbage (1791-1871), computer pioneer, designed the first automatic computing engines. He invented computers but failed to build them. The first complete Babbage Engine was completed in London in 2002, 153 years after it was designed. Difference Engine No. 2, built faithfully to the original drawings, consists of 8,000 parts, weighs five tons, and measures 11 feet long.

Big Bang

The hypothesis that our universe began approximately 14 billion years ago from a singularity—an ultra dense point containing everything we see today. In an ultra-brief moment in time it underwent rapid and enormous expansion, creating the atoms, all energy, and space-time which forms our inverse. It continues to inflate, driven by a unknown mechanism.

Black Hole

The idea of a body in space so massive that even light cannot escape from it. As light cannot escape, it would be seen as a black disk, hence the expression 'Black Hole'. If the semi-diameter of a sphere of the same density as the Sun were to exceed that of the Sun in the proportion of 500 to 1, a body falling towards it would acquire greater velocity than that of light. Consequently, any light which might be generated by the dense body would be attracted by the same force of gravity and would be pulled back in.

Branes

Branes are essentially membranes—lower-dimensional objects in a higher-dimensional space. (To picture this, think of a shower curtain, virtually a two-dimensional object in a three-dimensional space.) Branes are special, particularly in the context of string theory, because there's a natural mechanism to confine particles to the brane; thus not everything need travel in the extra dimensions even if those dimensions exist. Particles confined to the brane would have momentum and motion only along the brane, like water spots on the surface of your shower curtain. Branes allow for an entirely new set of possibilities in the physics of extra dimensions, because particles confined to the brane would look more or less as they would in a three-plus-one-dimension world; they never venture beyond it. Protons, electrons, quarks, all sorts of fundamental particles could be stuck on the brane.

Bubble Universe

This theory is when the universe grew exponentially in the first tiny fraction of a second after the Big Bang, some parts of space-time expanded more quickly than others. This could have created "bubbles" of space-time that then developed into other universes. The known universe has its own laws of physics, while other universes could have different laws, according to one interpretation of the multiverse concept.

C

Cyborg
A cyborg, short for "cybernetic organism", is a being with both biological and artificial (e.g. electronic, mechanical or robotic) parts. It may or not resemble a human being. When it does, it can also be referred to as an Android.

D

Dark Energy
Dark energy is a hypothetical form of energy which permeates all of space and tends to accelerate the expansion of the universe. It is the most accepted idea to explain the observations in space indicating that the universe is expanding at an accelerating rate. Many things about the nature of dark energy remain matters of speculation. It is thought that 68.3% of our universe is dark energy.

Dark Matter
Dark matter is a kind of matter in space which accounts for gravitational effects that appear to be the result of invisible mass. It cannot be seen directly with telescopes; evidently it neither emits nor absorbs light or other electromagnetic radiation at any significant level. The existence and properties of dark matter are inferred from its gravitational effects on visible matter, radiation, and the large-scale structure of the universe. It is considered that 26.8% of our universe is made up of Dark matter.

DNA
DNA (Deoxyribonucleic acid) is a molecule that encodes the genetic instructions used in the development and functioning of all known living organisms and many viruses. DNA is a nucleic acid. Alongside proteins and carbohydrates, nucleic acids compose the three major macromolecules essential for all known forms of life. Most DNA molecules consist of two biopolymer strands coiled around each other to form a double helix.

E

Entropy
Entropy is at root, a function of thermodynamic variables, as temperature, pressure, or composition, that is a measure of the energy that is not available for work during a thermodynamic process. A closed system evolves toward a state of maximum entropy. In cosmology, the area referred to within the scope of this book, it means a hypothetical tendency for the universe to attain a state of maximum homogeneity in which all matter is at a uniform temperature (heat death).

Event horizon
The gravitational sphere around a black hole within which the escape

velocity is greater than the speed of light, causing time to practically stop, and the point at which anything reaching the boundary of, is unable to escape. A threshold, where all the intuitive and logical perception of reality move away from a materialistic perspective and into one of metaphysical and often exotic hypothesis.

F

Fractals

A fractal is a never-ending pattern. Fractals are infinitely complex patterns that are self-similar across different scales. They are created by repeating a simple process over and over in an ongoing feedback loop. Driven by recursion, fractals are images of dynamic systems. Geometrically, they exist in between our familiar dimensions. Fractal patterns are extremely familiar, since nature is full of fractals. For instance: trees, rivers, coastlines, mountains, clouds, seashells, hurricanes, etc. Abstract fractals—such as the Mandelbrot Set—can be generated by a computer calculating a simple equation over and over.

G

Galaxy

A galaxy is a massive, gravitationally bound system consisting of stars, stellar remnants, an interstellar medium of gas and dust, and dark matter—a poorly understood component. Examples of galaxies range from dwarfs with as few as ten million stars to giants with one hundred trillion stars, each orbiting their galaxy's own centre of mass. A black hole is believed to be at the centre of most galaxies. Our own galaxy is called the 'Milky Way'.

Genes

A gene is the molecular unit of heredity of a living organism. It is used extensively by the scientific community as a name given to some stretches of deoxyribonucleic acids (DNA) and ribonucleic acids (RNA) that code for a polypeptide or for an RNA chain that has a function in the organism. Living beings depend on genes, as they specify all proteins and functional RNA chains. Genes hold the information to build and maintain an organism's cells and pass genetic traits to offspring.

GEO600

GEO600 is a ground-based interferometric gravitational wave detector located near Hannover, Germany. It is designed and operated by scientists from the Max Planck Institute for Gravitational Physics: **http://www.geo600.org/**

H

Hadron Collider
The Large Hadron Collider (LHC) is the world's largest and most powerful particle collider, built by the European Organization for Nuclear Research (known as CERN from the original French name "Conseil Européen pour la Recherche Nucléaire") from 1998 to 2008. It allows scientists to reproduce the conditions that existed within a billionth of a second after the Big Bang by colliding beams of high-energy protons or ions at colossal speeds, close to the speed of light. The LHC is exactly what its name suggests - a large collider of hadrons (any particle made up of quarks).

I

Inflation
Inflation holds that in the instant after the Big Bang start up of our universe, the initial point expanded rapidly—so rapidly that an area of space once a nanometre square ended up more than a quarter-billion light years across in just a trillionth of a trillionth of a trillionth of a second.

Ion
*See **Atom***

L

L.E.D. (Light Emitting Diode)
A light-emitting diode (LED) is a two-lead semiconductor light source, which emits light when a suitable voltage is applied to the leads. Electrons are able to recombine with electron holes within the device, releasing energy in the form of photons. This effect is called electroluminescence. They first appeared as practical electronic components in 1962.

M

Metaphysics
(From wiki)
Metaphysics is a traditional branch of philosophy concerned with explaining the fundamental nature of being and the world that encompasses it, although the term is not easily defined. Traditionally, metaphysics attempts to answer two basic questions in the broadest possible terms:What is ultimately there? What is it like?

A person who studies metaphysics is called a metaphysicist or a metaphysician. The metaphysician attempts to clarify the fundamental notions by which people understand the world, e.g., existence, objects and their properties, space and time, cause and effect, and possibility. A central branch of metaphysics is ontology, the investigation into the basic categories of being and how they relate to each other. Another central branch of metaphysics is cosmology, the study of the origin, fundamental structure, nature, and dynamics of the universe. Some include Epistemology as another central focus of metaphysics, but this can be questioned.

Prior to the modern history of science, scientific questions were addressed as a part of metaphysics known as natural philosophy. Originally, the term "science" (Latin scientia) simply means "knowledge". The scientific method, however, transformed natural philosophy into an empirical activity deriving from experiment unlike the rest of philosophy. By the end of the 18th century, it had begun to be called "science" to distinguish it from philosophy. Thereafter, metaphysics denoted philosophical enquiry of a non-empirical character into the nature of existence. Some philosophers of science, say that natural science rejects the study of metaphysics, while other philosophers of science strongly disagree.

Multiverse

The multiverse is the hypothetical set of infinite or finite possible universes (including the universe we consistently experience) that together comprise everything that exists: the entirety of space, time, matter, and energy as well as the physical laws and constants that describe them. The various universes within the multiverse are sometimes called parallel universes or "alternate universes".

N

Neuron

A neuron, also known as a neurone or nerve cell, is an electrically excitable cell that processes and transmits information through electrical and chemical signals. The human brain consists of billions of neurons interconnected by dendrites and synapses. Specialized types of neurons include: sensory neurons which respond to touch, sound, light and all other stimuli affecting the cells of the sensory organs that then send signals to the spinal cord and brain, motor neurons that receive signals from the brain and spinal cord to cause muscle contractions and affect glandular outputs, and inter-neurons which connect neurons to other neurons within the same region of the brain or spinal cord in neural networks.

P

Photon
Under the photon theory of light, a photon is a discrete bundle (or quantum) of electromagnetic (or light) energy. Photons are always in motion and, in a vacuum, have a constant speed of light to all observers, at the vacuum speed of light (more commonly just called the speed of light) of $c = 2.998 \times 10^8$ m/s. Light has properties of both a wave and a particle. Billiard balls act as particles, while oceans act as waves. Photons act as both a wave and a particle all the time (even though it's common, but basically incorrect, to say that it's "sometimes a wave and sometimes a particle" depending upon which features are more obvious at a given time).

Q

Quark
A quark is an elementary particle and a fundamental constituent of all matter. Quarks combine to form composite particles called hadrons, the most stable of which are protons and neutrons, the components of atomic nuclei. Due to a phenomenon known as colour confinement, quarks are never directly observed or found in isolation; they can be found only within hadrons, such as baryons (of which protons and neutrons are examples), and mesons. For this reason, much of what is known about quarks has been drawn from observations of the hadrons themselves. There are six types of quarks, known as flavours: up, down, strange, charm, bottom, and top.

Quantum Mechanics (Quantum Theory)
The word "quantum" in this sense means the minimum amount of any physical entity involved in an interaction. Certain characteristics of matter can take only discrete values. Quantum mechanics is the science of the very small: the body of scientific principles that explains the behaviour of matter and its interactions with energy on the scale of atoms and subatomic particles. Classical physics explains matter and energy on a scale familiar to human experience, including the behaviour of astronomical bodies. It remains the key to measurement for much of modern science and technology. However, toward the end of the 19th century, scientists discovered phenomena in both the large (macro) and the small (micro) worlds that classical physics could not explain. Quantum mechanics provides a mathematical description of much of the dual particle-like and wave-like behaviour and interactions of energy and matter.

Quantum Superposition
The principle of superposition states that if the world can be in any configuration, any possible arrangement of particles or fields, and if the world could also be in another configuration, then the world can also be in a state which is a superposition of the two, where the amount of each configuration that is in the superposition is specified by a complex number.

R

Robot
A robot is an automatic mechanical device often resembling a human or animal. Modern robots are usually an electro-mechanical machine guided by a computer program or electronic circuitry. A robot is a generic term - several different types of entities can be a robot. Androids and Cyborgs, for example, are also robots.

RNA
RNA (Ribonucleic acid) is a polymeric molecule. It is implicated in various biological roles in coding, decoding, regulating, and expressing genes. Like DNA, RNA is nucleic acids, and, along with proteins and carbohydrates, constitute the three major macromolecules essential for all known forms of life. RNA is assembled as a chain of nucleotides, but unlike DNA it is more often found in nature as a single strand folded unto itself, rather than a paired double-strand.

S

Singularity
It is better called a gravitational singularity, a one-dimensional point which contains infinite mass in an infinitely small space, where gravity become infinite and space-time curves infinitely, and where the laws of physics as we know them cease to operate. An example is a 'Black Hole'.

String Theory
In the last few decades, string theory has emerged as the most promising candidate for a microscopic theory of gravity. And it is infinitely more ambitious than that: it attempts to provide a complete, unified, and consistent description of the fundamental structure of our universe. (For this reason it is sometimes, quite arrogantly, called a 'Theory of Everything'). The idea behind string theory is this: all of the different 'fundamental ' particles are really just different manifestations of one basic object: a string. We would ordinarily picture an electron, for instance, as a point with no internal structure. A point cannot do anything but move. But, if string theory is correct, then under an extremely powerful hypothetical 'microscope' we would realize that the electron is not really a point, but a tiny loop of string. A string can do something aside from moving---it can oscillate in different ways. If it oscillates a certain way, then from a distance, unable to tell it is really a string, we see an electron. But if it oscillates some other way, well, then we call it a photon, or a quark, etc. So, if string theory is correct, the entire world is made of strings!

U

UFOs

UFO is an abbreviation for Unidentified Flying Objects. In its most general definition, it is any apparent anomaly in the sky that is not identifiable as a known object or phenomenon. Such anomalies may later be identified, but depending on the evidence or lack of evidence, such an identification may not be possible, generally leaving the anomaly unexplained. While stories of unexplained apparitions have been told since antiquity, the term "UFO" was officially created in 1953 by the United States Air Force (USAF) to serve as a catch-all for all such reports. The term is often used more loosely as a reference to 'flying saucers' or Ariel craft presumed to be of alien origin..

Unitarity

In quantum physics, Unitarity is a restriction on the allowed evolution of quantum systems that ensures the sum of probabilities of all possible outcomes of any event is always 1. More precisely, the operator which describes the progress of a physical system in time must be a unitary operator.

W

Wave Function

A wave function is referred to in the field and research of quantum mechanics. It describes the quantum state of a system of one or more atomic or sub-atomic particles, and contains all the information about the system considered in isolation. Quantities associated with measurements, such as the average momentum of a particle, are derived from the wave function by mathematical operations describing its interaction with observational devices. It is a mathematical tool and not thought to be a physical characteristic of particles themselves.

APPENDICES

These include references to books and web links for anyone who might wish to follow up any of the statements made. They include some of my research material with additional works or internet-presented material not necessarily looked at in a detailed way, but chosen as a fair balance of *considered* material. My research has been over many years and not part of a focused academic pursuit. I cannot include everything which has influenced my final thoughts and position regarding the thrust of, and direction of, this book. I simply would not remember all the influences.

The quantity of information available today due to the internet is beyond usefulness. Much of it is pitched at a level too high or too low, or it is often too diluted, or exploited by 'lone rangers' to create boring and non-informative Youtube videos.

I have selected those references which might best invoke further interest in anyone who is serious about considering the issues in my book.

Also, a small work like mine cannot possibly describe the 'big picture'. The reader should see it as a 'road sign' to a vaster story which conventional science is not really sign-posting. I have attempted to filter out anything too far removed from real clues and potential evidence. I also stayed away from anything I believed to be exploited and perverted to any kind of dogmatic religious view. Sadly, any/all knowledge and ideas seem to be seized upon and twisted to fit biased idealisms. I might have missed one or two in my selection, or included one I meant to avoid. I hope not. But if I have included a reference from a 'loony', it will be an error. Forgive me for that if I did, and just move on.

Many very respected scientists, thinkers, philosophers, physicists, doctors, researchers and intelligent rational people support a view of the world and our reality which is far removed from the mechanical, materialistic perspective of mainstream science. They are bold and pioneering thinkers. I hope you will encounter them through these references.

Appendix 1: Book & Internet References
Web links can also be found at: *www.42plus1.net*

Chapter 1: From then to now

Books

Web Links for Chapter 1 follow overleaf!

Worlds In Collision
Re-Publisher: Paradigma Ltd (Oct 2009)
ISBN-10: 1906833117
ISBN-13: 978-1906833114

Earth In Upheaval
Publisher: Abacus (1976)
ISBN-10: 0349135762
ISBN-13: 978-0349135762

Ages in Chaos
Publisher: Abacus Books (Sphere) (Dec 1973)
ISBN-10: 0349135827
ISBN-13: 978-0349135885

Author: Immanuel Velikovsky
1895-1979
http://www.varchive.org/

Books exploring a different history of earth often based on biblical text and Jewish texts and events mentioned in them. Although the events themselves, which involved things like planetary collisions, are still refuted today—the notion of earth being shaped and affected by chaotic and catastrophic events was introduced and now accepted as true. It's just the details of what caused those events is not accepted. Before the 1950's the idea of large mass meteor strikes and super volcanoes were not widely accepted scientific theory.

Shadows Of The Mind
Publisher: Vintage; New Ed edition (7 Sep 1995)
ISBN-10: 0099582112
ISBN-13: 978-0099582113

Author: Roger Penrose

A profound exploration of what modern physics has to tell us about the mind, and a visionary description of what a new physics - one that is adequate to account for our extraordinary brain - might look like.

Appendix 1: Book & Internet References
Web links can also be found at: *www.42plus1.net*

Chapter 1: From then to now
Web Links. These can also be found at: ***www.42plus1.net***

Amoeba — http://en.wikipedia.org/wiki/Amoeba
Video — https://www.youtube.com/watch?v=7pR7TNzJ_pA

Big Bang Theory — http://science.nationalgeographic.com/science/space/universe/origins-universe-article/
http://en.wikipedia.org/wiki/Big_Bang
https://www.youtube.com/watch?v=MfpH3Zox6m4

Bubble Universes *(also see **Multiverse**)* — http://www.perimeterinstitute.ca/news/universe-bubble-lets-check
Video — https://www.youtube.com/watch?v=MfpH3Zox6m4

Multiverse — http://en.wikipedia.org/wiki/Multiverse

Carbon Based Life — http://en.wikipedia.org/wiki/Carbon-based_life

Carbon — http://en.wikipedia.org/wiki/Carbon
http://www.edinformatics.com/math_science/c_atom.htm

Paramecium — http://101science.com/paramecium.htm
Video — https://www.youtube.com/watch?v=zS0f82ZJtvk

Roger Penrose — www.maths.ox.ac.uk/people/roger.penrose [contact]
Video — http://www.usfca.edu/artsci/math/penrose-2014/

Chapter 2: Reality is obscure
Web Links. These can also be found at: ***www.42plus1.net***

Double Slit Experiment
About — http://en.wikipedia.org/wiki/Double-slit_experiment
Video — https://www.youtube.com/watch?v=fwXQjRBLwsQ

Quantum Entanglement
About — http://en.wikipedia.org/wiki/Quantum_entanglement
Video — https://www.youtube.com/watch?v=ZNedBrG9E90

Appendix 1: Book & Internet References
Web links can also be found at: *www.42plus1.net*

Chapter 3: Many worlds: multiple realities
Web Links. These can also be found at: ***www.42plus1.net***

Hugh Everett III (November 11, 1930 – July 19, 1982)	http://www.philosophy-of-cosmology.ox.ac.uk/everett.html
Many Worlds————————	http://en.wikipedia.org/wiki/Many-worlds_interpretation
Wave Function	
About————————	https://en.wikipedia.org/wiki/Wave_function
Video ————————	https://www.youtube.com/watch?v=Ei8CFin00PY

Chapter 4: Who am I?
Web Links. These can also be found at: ***www.42plus1.net***

Neurons	http://en.wikipedia.org/wiki/Neuron
Reincarnation	
The Boy Who Told Who Killed Him (Book on Amazon)	http://www.amazon.co.uk/Children-Have-Lived-Before-Reincarnation/dp/1844132986/
Old Souls (Dr. Ian Stevenson) Book by Tom Shroder On Amazon	http://www.amazon.co.uk/Old-Souls-Scientific-Evidence-Search/dp/0684851938
Spatial Awareness	http://www.youtube.com/watch?v=mD7NzrBgXwM

Chapter 5: Clues in our biology point to our profound identities
Web Links. These can also be found at: ***www.42plus1.net***

Author: Rupert Sheldrake	http://www.sheldrake.org/
Science Set Free Video	https://www.youtube.com/watch?v=UPccMlgug8A
Memory Fields Lecture (Audio)	http://www.sheldrake.org/audios/memory-morphic-resonance-and-the-collective-unconscious
Book: The Science Delusion	http://www.amazon.co.uk/The-Science-Delusion-Rupert-Sheldrake/dp/144472794X

Appendix 1: Book & Internet References
Web links can also be found at: *www.42plus1.net*

Chapter 6: Why is there anything?
Web Links. These can also be found at: ***www.42plus1.net***

Big Bang Video	http://www.bbc.co.uk/science/space/universe/questions_and_ideas/big_bang/#p009fkpj
Dark Matter	http://en.wikipedia.org/wiki/Dark_matter
Vacuum Fluctuation	
About————————	http://en.wikipedia.org/wiki/Quantum_fluctuation

---

Chapter 7: A miracle?
Web Links. These can also be found at: ***www.42plus1.net***

Drawings Of Algae	
Book by Christine Brodie	http://www.amazon.co.uk/Drawing-Painting-Plants-Christina-Brodie/dp/071366889X
Emergence of life (Wiki)	http://en.wikipedia.org/wiki/Abiogenesis
Fractals	http://fractalfoundation.org/resources/what-are-fractals/
Goldilocks Universe	http://en.wikipedia.org/wiki/Fine-tuned_Universe
Stephen Wolfram (Web site)	http://www.stephenwolfram.com/
Book: A new kind of science	http://www.amazon.co.uk/New-Kind-Science-Stephen-Wolfram/dp/1579550088

Chapter 8: One mind: many minds?
Web Links. These can also be found at: ***www.42plus1.net***

DMT: The God drug (wiki)	http://topdocumentaryfilms.com/dmt-the-spirit-molecule/
Video————————	https://www.youtube.com/watch?v=EZAMKn2xr9E
Neutrinos	http://en.wikipedia.org/wiki/Neutrino
NDE (Near death experience)	
Video————————	https://www.youtube.com/watch?v=M4PmjKn1zPE

Appendix 1: Book & Internet References
Web links can also be found at: *www.42plus1.net*

Chapter 9: Where do we live now?
Web Links. These can also be found at: ***www.42plus1.net***

Most resolved image of the Milky Way	http://www.eso.org/public/images/eso1242a/zoomable/
NASA Images	http://www.nasa.gov/multimedia/imagegallery/index.html#.VKqgiSusV8E

Chapter 10: Rise of the machines
Web Links. These can also be found at: ***www.42plus1.net***

Asimo Robot (specs)	http://asimo.honda.com/downloads/pdf/asimo-technical-information.pdf
Video————————	https://www.youtube.com/watch?v=eU93VmFyZbg
Web Site———————	http://asimo.honda.com/
Female Humanoid Robot	http://global.kawada.jp/mechatronics/hrp4.html
Video————————	https://www.youtube.com/watch?v=jYvbjaDWHAg
Swarming Drones	
Grasp Web Site——————	https://www.grasp.upenn.edu/
Video————————	https://www.youtube.com/watch?v=YQIMGV5vtd4#t=49

Chapter 11: Purpose
Web Links. These can also be found at: ***www.42plus1.net***

No external references

Chapter 12: Are we really somewhere else?
Web Links. These can also be found at: ***www.42plus1.net***

Black Holes (About)	http://en.wikipedia.org/wiki/Black_hole
Event Horizon	http://en.wikipedia.org/wiki/Event_horizon
David Bohm (Web Site)	http://dbohm.com
About————————	http://en.wikipedia.org/wiki/David_Bohm
Video————————	https://www.youtube.com/watch?v=QI66ZglzcO0
Holometer (Fermilab)	http://holometer.fnal.gov/

Appendix 1: Book & Internet References
Web links can also be found at: *www.42plus1.net*

Chapter 13: Conclusions
Web Links. These can also be found at: ***www.42plus1.net***
No external references.

Chapter 14: Spiritual Considerations?
Web Links. These can also be found at: ***www.42plus1.net***
No external references.

Chapter 15: Message in a bottle
Web Links. These can also be found at: ***www.42plus1.net***
No external references

NOTES
(In case you wish to make some)

THE END

(Never happens...)

www.ingramcontent.com/pod-product-compliance
Lightning Source LLC
Chambersburg PA
CBHW051653170526
45167CB00001B/455

* 9 7 8 1 5 0 6 0 2 2 8 9 5 *